IN SACRED RELATIONSHIP
A Spiritual Compass for Today's Turbulent Times
Inspired by Lakota Wisdom

Donald T. Iannone
D. Div., M. Div., M.A.

Dear Carol,

Hope you find your own personal spiritual compass in this book.

~Don
11/16/2020

In Sacred Relationship

Copyright © 2020 Donald T. Iannone

All rights reserved.

DEDICATION

In Sacred Relationship is dedicated to the Lakota people, who generously share their spiritual wisdom with others, including non-Indians, who are in search of truth, healing, and wholeness. *Wacantognaka* is the Lakota word for generosity, which means to contribute to the well-being of one's people and all life by sharing and giving freely. I am grateful for the Lakota way of *Wacantognaka.* May we all be so generous with our wisdom and love.

CONTENTS

Dedication

Acknowledgments

Part I: Introduction and Orientation to *Book*   1

Chapter 1: Introduction   2

Part II: Lakota Spirituality and Sacred Land   19

Chapter 2: Overview of Indian Tribes Sharing Geography
with South Dakota   20

Chapter 3: Overview of Lakota History, Culture, and Spirituality   26

Chapter 4: Understanding Sacred Land   43

Chapter 5: The Black Hills, Badlands, and Devils Tower
as Lakota Sacred Lands   51

Chapter 6: Lakota Spiritual Worldview and Sacred Land   68

**Part III: Working with Grounded Spirituality and
Sacred Relationship (GSSR)**   76

Chapter 7: The Grounded Spirituality and Sacred
Relationships (GSSR) Compass   77

Chapter 8: Three Applications of the GSSR: Cancer Care,
The Coronavirus, and Serving Others in Sacred Relationship   128

Chapter 9: Concluding Thoughts   149

About the Author   151

References   152

Appendices   162

Index   179

In Sacred Relationship

ACKNOWLEDGMENTS

I would like to thank everyone who contributed to In Sacred Relationship by sharing their knowledge, ideas, and experiences. In particular, I wish to thank Professor Jace DeCory (Cheyenne River Lakota Nation and Black Hills State University), Roger Broar (Oglala Lakota Nation and Roger Broar Fine Arts), Professor David Posthumus (Anthropology and Sociology, University of South Dakota), Professor Larry Zimmerman (Anthropology and Museum Studies, Indiana University/Purdue University at Indianapolis), Professor Margaret Branch (University of Metaphysical Sciences), Rev. Josanne Pagel (Cleveland Clinic), David Michael Kennedy (American Indian photographer, David Michael Kennedy Photography), and Stephen Buttress (Retired Economic and Community Development Executive and Author). Finally, my wife Mary Iannone has been incredibly supportive of my writing this book. She participated in the fieldwork for this book in South Dakota and Wyoming last fall.

PART I:
INTRODUCTION AND ORIENTATION TO
IN SACRED RELATIONSHIP

"When you know who you are; when your mission is clear, and you burn with the inner fire of unbreakable will; no cold can touch your heart; no deluge can dampen your purpose. You know that you are alive."
~ Chief Seattle, Duwamish Chief.

CHAPTER 1: INTRODUCTION

We need a spirituality that grounds us in our relationships with ourselves, others, and the whole of life. We need a spirituality that helps us relate in a sacred way to the vast and everchanging web of life. We need a spirituality that helps us see the sacred in everyone and everything. We need a spirituality we can live each day that brings us balance, peace, harmony, and wholeness. This is the Lakota Way. This is grounded spirituality.
Don Iannone

A. An Orientation to This Book

Spiritual Growth through Relationships

In Sacred Relationship explains how we can grow spiritually by changing how we relate to the world. The first third of the book orients us to the book. The middle third of this book explores Lakota spirituality and examines its significance. The final third of the book uses ideas from Lakota spirituality to create a new spiritual compass model that helps people navigate more successfully in today's turbulent and uncertain world. The compass is a tool for use in any religion or spiritual tradition.

This book is about spiritual realities, which have been informed by scientific research. While science can help some in explaining spiritual realities in the context of human brain functions and processes, spiritual *"experiences"* remain largely unexplained by science. For example, we know that scientific research in neurotheology is helping us to understand the role of the brain in religious experience.[1][2][3][4][5] Scientific research also informs us that patient health is improved by spiritual belief and practice.[6]

---

[1] Newberg, Andrew B. Principles of neurotheology. Ashgate Publishing, Ltd., 2010.
[2] Newberg, Andrew. "How God changes your brain: An introduction to Jewish neurotheology." CCAR Journal: The Reform Jewish Quarterly, Winter (2016).
[3] Ashbrook, James B. "Neurotheology: the working brain and the work of theology." Zygon® 19.3 (1984): 331.
[4] Harper, Kate. "Historical and Contemporary Explorations in Neurotheology." Consensus 40.2 (2019): 4.
[5] Clark, Alexis Elizabeth. "A literature review of neurotheology: How religion affects the brain." (2018).
[6] Steinhorn, David M., Jana Din, and Angela Johnson. "Healing, spirituality and integrative medicine." Annals of palliative medicine 6.3 (2017): 237-247.

[7] [8] [9] Some researchers attribute this health improvement to the placebo effect of spirituality, which in this case is defined as a beneficial health effect produced by spirituality as a placebo, and the patent's belief in the value of the placebo (spirituality) in health improvement.[10] [11] [12] [13] The placebo effect testifies to the power of belief in the human experience. Quite importantly, the lack of a robust scientific explanation of religious and spiritual experience does <u>not</u> negate the important role played by religion and spirituality, and their meaningfulness to all of us.

While the scientific explanation of spiritual experience is not the topic of this book, the existing scientific research on spiritual experience has been examined, and it was considered in the development of the spiritual compass presented in the book's second half. Science and spirituality need each other to provide a more complete understanding of the human quest to find meaning, purpose, and wholeness in life.

How to View Sacred Relationships

Sacred relationships are best viewed as relationships within spiritual reality, which help people to *"promote and experience the sacred." "Sacred"* means to flow from the Divine or God, to be dedicated to spiritual purpose, and to be worthy of veneration. This book takes the whole of life to be sacred, including our relationships, because life is a Divine gift animating our world. This idea sits at the center of Lakota spirituality, and *In Sacred Relationship* builds upon this idea.

Reader Audiences

In Sacred Relationship was written for readers looking for a grounded way to energize their spiritual or religious faith through *sacred relationship-building*, which is an idea I first encountered in Lakota spirituality. The underlying premise of this book is that we can increase

---

[7] Mishra, Shri K., et al. "Spirituality and religiosity and its role in health and diseases." Journal of religion and health 56.4 (2017): 1282-1301.
[8] Torosian, Michael H., and Veruschka R. Biddle. "Spirituality and healing." Seminars in oncology. Vol. 32. No. 2. WB Saunders, 2005.
[9] Jonas, Wayne B., and Cindy C. Crawford. "A critical review of spiritual healing," Alternative therapies in health and medicine 9 (2003): 2.
[10] Puchalski, Christina M. "The role of spirituality in health care." Baylor University Medical Center Proceedings. Vol. 14. No. 4. Taylor & Francis, 2001.
[11] Alling, Frederic A. "The healing effects of belief in medical practices and spirituality." Explore 11.4 (2015)
[12] Mueller, Paul S., David J. Plevak, and Teresa A. Rummans. "Religious involvement, spirituality, and medicine: implications for clinical practice." Mayo clinic proceedings. Vol. 76. No. 12. Elsevier, 2001.
[13] Harrington, Anne. "Placebo effect: What's interesting for scholars of religion." Zygon® 46.2 (2011): 265-280.

sacredness and goodness in daily lives by changing how we relate to God or the Divine, our minds, bodies, and spirits, other people, and the whole of life. The *new spiritual compass* presented in this book is designed to help each of us within our religion or spiritual faith.

Because of the extensive research underlying this book, *In Sacred Relationship* is suitable as an academic text and professional training guide for students studying in fields such as religion and spirituality, philosophy and metaphysics, spirituality in medicine, nursing, psychology, and social work, spiritual care and counseling, and the anthropology and sociology of religion.

In addition, *In Sacred Relationship* offers wisdom to business, community, healthcare, education, and government leaders on how to relate in a more sacred way to their work and those they serve. A central challenge for leaders in all fields is to add greater value to customers, clients, patients, students, and citizens. *In Sacred Relationship* can help leaders achieve this by honoring the sacred in their work relationships with others. *This book promotes the idea that sacred relationships can give rise to a new spiritually centered model for engaging and serving patients, customers, clients, citizens, and students.*

In Sacred Relationship is for people of all faiths, and people with no declared religious faith. New research calls attention to the perceived need for religion and spirituality to play a greater role in everyday life affairs. This research finds that a <u>declining</u> percentage of Americans believe their religion is very important in their lives. In 2003, 61 percent of Americans responding to the Gallup Religion Poll said their religion was very important in their lives, compared to only 49 percent in 2019.[14] This is a 12 percent drop. While many Americans remain highly committed to their faith, a large number of Americans have *"lost faith in their faith"* because they do not believe it offers them enough in terms of *everyday value.* Seventy-five percent of Americans polled by Gallup in 2019 said that connecting their religion to their personal lives in more meaningful ways was a major priority, but unfortunately their religion is not doing enough to achieve this priority.[15] *In Sacred Relationship* aims to help us *live our faith* with greater practicality, relevance, and impact.

---

[14] Gallup Religion Poll, March 2019: <u>https://news.gallup.com/poll/1690/religion.aspx</u>. Access January 20,2020.
[15] Ibid.

B. Why I Really Wrote This Book

I wrote this book because life is sacred, and today's world is broken and torn because of our individual and collective inabilities to relate in sacred ways to ourselves, others, and the whole of life. *The Coronavirus Pandemic is a frightening and sad reality in its own right, and it is also a clarion call about how we have fallen out of sacred relationship with ourselves and our world.* If we use this crisis in the right way, it can help us develop greater compassion, balance, and responsibility by changing our relationship to the world. We need a new spiritual compass to help us find our way back to the sacred in everyday life. In the absence of sacredness, our lives grow empty, superficial, and overly materialistic. America's consumer culture is a result of the loss of sacred meaning in our daily lives. Our consumer culture has deceived us into believing we can buy true happiness, peace, and meaning, and it has led many to approach spirituality as consumers, rather than seeing it as a source of vital sustenance for our natural divinity.

Our determination to thrive and flourish has been overthrown by our frightening desperation to survive through conquest, manipulation, and truth-avoidance. We look to football and other sports to feed our spirits with competitive rivalry and zero-sum winning, and yet we fail to engage in meaningful teamwork with our co-workers on a daily basis. We idolize wealthy athletes, financiers, and technology entrepreneurs, and fail to give a second thought to the millions of people who barely survive each day in American society. In our supposed *"Age of Talent,"* we train MBAs, doctors, lawyers, actors, investment bankers, and politicians to become narcissistic celebrity heroes in today's personally branded and individualistic economy. We are losing our ability to think critically and feel genuinely about important life issues, favoring instead the information bytes fed to us by the print and electronic news bureaus and social media.

Is this an over-dramatization of the current state of affairs in our world today? I do not believe so. Outwardly, we say these things are *"the new normal,"* while inwardly we are torn to pieces by our spiritual bankruptcy. *In Sacred Relationship* provides an authentic way out of this downward spiral. *Let's remember and relearn the lesson that life is a precious gift, which we should regard as sacred ground in all we do. Our relationships define us. If we relate to life as a sacred gift, a new world of wholeness and healing appears.* That is the vision guiding this book.

In Sacred Relationship

As Lakota spirituality teaches, *sacred ground is common ground,* which we share with others, allowing us to grow together in goodness. Our world today is locked in destructive ideological conflicts about politics, religion, race, and culture, which have divided us, and spurred selfishness, hatred, and fear, causing us to lose our way in finding peace and truth in our lives. These realities have bred a deep unsettledness inside each of us and in the world surrounding us. This unsettledness undermines our goodness, sense of hope, and ability to care about ourselves, others, and the world. *We are larger than all this because divinity is our true nature. In Sacred Relationship* offers us a way back to our natural divinity!

When I look at current events in our Nation's capital, in America's big cities and small towns, on American Indian reservations, and elsewhere around the world, I am dismayed to see that we have forsaken the precious gift of life itself. Our elected leaders are in constant warfare with each other, leaving them little time to serve and protect citizens. These leaders have ignored science (what we know is true) and ethics (what we know is right) in peeling back important regulations protecting our natural environment.

The people's commonwealth (our shared public goods, including our sacred lands) has been (mis)appropriated for personal gain and power by our elected officials, and those who buy elections for them. Many large global corporations conduct themselves like barbarous nation-states, hoarding wealth and power, and dismissing their responsibilities to workers, communities, and the natural environment. Distraught and disturbed adults and children use automatic weapons to massacre dozens of classmates, work colleagues, fellow worshippers, and complete strangers because of their own brokenness, and because life to them is no longer sacred. Finally, members of the Church itself have in many cases forsaken their followers by stealing, misleading, and abusing their congregations, who look to them to set examples of goodness and provide them with meaningful spiritual care and hope.[16] [17] [18]

---

[16] Flynn, Meagan, A disgraced televangelist promoted an alleged cure to coronavirus. Missouri is now suing him, Washington Post, March 11, 2020: https://www.washingtonpost.com/nation/2020/03/11/jim-bakker-coronavirus-cure/. Accessed March 13, 2020.

[17] Pongratz-Lippitt, Christa, Martha Pskowski, Abuse scandals could 'destroy' Church, The Tablet, February 22, 2020: https://www.thetablet.co.uk/news/12520/abuse-scandals-could-destroy-church. Accessed March 13, 2020.

[18] Dias, Elizabeth, Her Evangelical Megachurch Was Her World. Then Her Daughter Said She Was Molested by a Minister, New York Times, August 10, 2019: https://www.nytimes.com/2019/06/10/us/southern-baptist-convention-sex-abuse.html

These issues dominate our news headlines on a daily basis. Many of us are deeply troubled by these dire realities, and we are struggling to find effective responses to these realities. We have forgotten how to honor the sacredness of life. It does not matter if you are Liberal, Conservative, Christian, Islamic, Buddhist, Agnostic, White, Black, Red, Yellow, financially rich or poor, these are deep-seated realities that are destroying our world because we fail to honor life's sacredness on our Planet. This book offers us new hope, which says that by changing how we relate to ourselves and the world, we can bring our lives and the world into better balance.

Life is a sacred circle, as the Lakota remind us, which means that eventually there is no escape from the disastrous consequences of our current worldviews and actions. We have forgotten that we <u>only</u> exist in relation to each other. Our relationships define us. When our relationships are broken, we are broken. Moreover, our universal bond of divinity has been shattered, and traded away in uncaring supply and demand relationships. These conditions *reflect an even deeper and more frightening problem in American society: On a personal level, we have fallen out of sacred relationship balance with ourselves and our world. These are not just words; they are realities we awaken to each day. These are the reasons why In Sacred Relationship is important.*

With regard to the views expressed in *In Sacred Relationship*, I am very aware of the concern by American Indians about non-Indians seeking to explain their history and spirituality. My experience in working with American Indian tribes and academic training in Anthropology have made me aware of these concerns. I am grateful for the contributions provided to my work by other researchers and members of the Lakota Nation. Their help has added greatly to this book's credibility, and yet what I have written about Lakota spirituality remains an *"outsider perspective."* In this book, I use what I have learned from the Lakota to formulate *my own truth*, which I have spoken throughout this book. As the reader, I invite you to do the same.

C. Book Contents

Four reasons explain why this book gives major attention to Native American spirituality, and especially Lakota spirituality. First, they have spoken to my heart and reminded me of life's inherent sacredness.

Second, I wrote my Master of Divinity thesis on Lakota spirituality and sacred land, and I wanted to share the best part of that research in a relevant way with others. Third, Native American spiritual traditions encourage us to live in balance and harmony, which are essential ingredients to spiritual growth and personal healing. Finally, Lakota spirituality is a *spiritually grounded way of life*, which is greatly needed in today's uprooted and ungrounded world.

With respect to the Lakota spiritual content of the book, historical and contemporary perspectives of the Lakota spirituality are included in this book to avoid *"romanticizing"* the American Indian world of long ago. This contemporary perspective ensures that current issues and realities facing the Lakota people are reflected in the book. While a traditional, or historically based understanding of Lakota spirituality and sacred land is valuable, it falls short in understanding the role of religion, spirituality, and sacred land in today's American Indian world. The world changes, and religion and spirituality also change as integral aspects of the world.

A grounded approach to spirituality and religion helps us to keep in step with our rapidly changing world. To do that, spirituality and science must work together, as the astrophysicist and cosmologist Carl Sagan reminds us: *"The notion that science and spirituality are somehow mutually exclusive does a disservice to both."*[19] While this book focuses on life's spiritual realities, it also honors science. Hopefully, science and spirituality will someday build a sacred relationship with each other in bringing the integrated wisdom, knowledge, and understanding that the world needs.

The centerpiece of this book is how I use concepts from Lakota spirituality to build a new spiritual compass, which I have named the Grounded Spirituality and Sacred Relationship Compass (GSSRC)©. The compass is my way of offering a spiritual growth and improvement tool that people of all faiths can use.

In Sacred Relationship has three parts, which are organized into nine chapters. The first part of the book orients the reader to the Lakota world with short sections on tribal demographics, history, and culture. It also includes a section about my personal interest in the Indian world, and why I authored this book. I discuss how my early experiences with the Native American world shaped my spiritual curiosity. I believe we should learn something from books we write. The writing of this book has been a

---

[19] Sagan, Carl. The demon-haunted world: Science as a candle in the dark. Ballantine Books, 2011.

valuable opportunity for me to explore and strengthen my spiritual life.

The second part of the book explores Lakota spirituality and the role of sacred land in the Lakota spiritual worldview. This part of the book discusses the key Lakota spirituality themes in a condensed way. Many in-depth books already exist on Lakota spirituality. [20] [21] [22] [23] [24] [25] [26] [27] These books focus on the important teachings of Lakota traditional healers and elders, such as Black Elk and Luther Standing Bear. Other books offer perspectives on how Lakota spirituality relates to Lakota culture, healing, and medicine.

Real World GSSR Compass Test Applications:

The grounded spirituality and sacred relationship compass (GSSRC) is applied in this book to:

1. The care of cancer patients and their families;
2. The novel coronavirus public health threat; and
3. Customer, client, and patient engagement and service in several fields.

The book's third part discusses how Lakota spirituality leads us to *"sacred relationships"* and *"grounded spirituality."* In short, grounded spirituality embodies spiritual beliefs and practices that ground us in the everyday world, helping us to see the beauty and specialness in our everyday lives. Grounded spirituality is practical spirituality that encourages us to seek balance, harmony, and wholeness in our lives. Part 3 helps us understand Lakota spirituality as a foundation for grounded spirituality and sacred relationship-building. The GSSRC guides us in living *"in sacred relationship"* with our world.

In the last section of Part 3, the GSSRC is applied to three important areas of concern in today's world to test its utility and value. The first application is in serving the spiritual needs of cancer patients. My work with cancer patients indicates that patients and their families are open to new sources of spiritual wisdom and care.

The second is a test of the compass's value as a strategy to create

---

[20] Posthumus, David C. All My Relatives: Exploring Lakota Ontology, Belief, Ritual. U of Nebraska Press, 2018.

[21] Flood, Reneé S. Lost Bird of Wounded Knee: Spirit of the Lakota. NY: Scribner, 1995.

[22] LaPointe, James, and Louis Amiotte. Legends of the Lakota. Indian Historian Press, 1976.

[23] Elk, Wallace Black, and William S. Lyon. Black Elk: The sacred ways of a Lakota. HarperCollins Pub., 1990.

[24] Ellerby, Jonathan H. "Spirituality, holism and healing among the Lakota Sioux, towards an understanding of Indigenous medicine." (2000).

[25] Arden, Harvey, and Steve Wall. Wisdomkeepers: Meetings with Native American Spiritual Elders. Beyond Words Publishing, Inc., 4443 NE Airport Rd., Hillsboro, OR

[26] Walker, James R. Lakota belief and ritual. U of Nebraska Press, 1980.

[27] Erdoes, Richard. Crow Dog: Four generations of Sioux traditional healers. New York: HarperCollins, 1995.

awareness of the spiritual realities associated with the new coronavirus public health threat.[28] Spiritual realities surround the new coronavirus, in addition to its important scientific, medical, and economic realities. The GSSRC can help us learn important spiritual lessons from this public health threat, meet the spiritual needs of those touched by this disease, and cope with the virus in more caring, collaborative, and effective ways. Honoring life globally is more possible when we see the sacred or divine in all people and things. The GSSRC helps us do this. In the third area, it shows how the compass can give impetus to an improved approach to engaging and serving patients, students, customers, clients, and citizens. By cultivating sacred relationships with those we serve, we honor life's sacredness in our service to others.

D. Book Guiding Purposes and Objectives

Guiding Purposes

This book has three purposes, which correspond to the three main book parts. The book's first purpose is to provide the reader with an understanding of Lakota spirituality.[29] The second purpose is to share a new spiritual compass; which readers can use in practical ways to strengthen their own religion or spirituality. The third purpose is to provide initial tests of the compass in three important areas to illustrate how it may be beneficial.

Book Guiding Purposes:
1. Understand Lakota spirituality;
2. Present a model (compass) for grounded spirituality and sacred relationships for today's world; and
3. Illustrate the compass's usefulness through application to cancer care, the novel (new) coronavirus threat situation, and customer/client and patient engagement and service.

Objectives

Five specific objectives guided the writing of this book. First, the Lakota spiritual worldview is

---

[28] Much is unknown about how 2019-nCoV, a new coronavirus, spreads. Current knowledge is largely based on what is known about similar coronaviruses. Coronaviruses are a large family of viruses that are common in many different species of animals, including camels, cattle, cats, and bats. Rarely, animal coronaviruses can infect people and then spread between people such as with MERS, SARS, and now with 2019-nCoV. Source: Center for Disease Control: https://www.cdc.gov/coronavirus/2019-ncov/about/transmission.html. Accessed February 16, 2020.

[29] The term "Lakota" is used to refer to the seven bands of the Tetonwan, or the Teton Lakota group. Within this group, greatest attention is given to the Oglala Lakota, who primarily reside on and in the area surrounding the Pine Ridge Reservation in southwestern South Dakota. The "Sioux" name, which is used less often today in referring to the Lakota people, is a reference to the Great Sioux Nation, or the Očhéthi Šakówiŋ, which is translated as the "Seven Council Fires" or the seven bands.

explored and discussed in terms of its uniqueness, and what it shares in common with other indigenous spiritual traditions. The second aim is to provide an understanding of the Native American concept of sacred land, especially from the Lakota perspective. The third objective is to examine the Black Hills and the Badlands in western South Dakota and Devils Tower in eastern Wyoming as natural areas that are considered to be sacred land by the Lakota people.[30] [31] Fourth, the relationship between Lakota spirituality and sacred land is examined and discussed. The three-part issue explored in this fourth objective is: 1) how geography and place have shaped the Lakota spiritual worldview; 2) how the Lakota spiritual worldview has contributed to the Lakota view of and interaction with specific geographies; and 3) how sacred meaning is attached to specific places by the Lakota. The book's final aim is to show how Lakota spirituality can help us find spiritual grounding in today's world. This is achieved through the GSSR Compass.

In the book's last chapter, the Grounded Spirituality and Sacred Relationship Compass (GSSRC) is applied to the three areas of concern identified earlier. These first tests find the compass to be very relevant and beneficial in all three applications. Also, they point to future work that is needed in developing the compass as a more highly useable tool.

E. My Personal Spiritual History

To fully understand what this book is about, I believe that I owe the reader some explanation of my own personal spiritual path, which has influenced this book's content and approach. My original spiritual grounding was in the faith that today we call Evangelical Christianity. For many years, my family belonged to the Nazarene Church, which in the 1950s and 1960s adhered to a very strict and literal view of the Bible and the teachings of Jesus Christ. The Nazarene Church taught its followers to walk steep and narrow paths of righteousness, sidestepping *worldliness* in life in all forms. The trouble with this path is that our very nature is constantly shaped by our worldly and divine instincts. Our lives emerge in the dialectical dance between the two. We discover divinity through our

---

[30] Clark, Linda, Darus, Sioux Treaty of 1868, National Archives, at: https://www.archives.gov/education/lessons/sioux-treaty, accessed August 31, 2019.
[31] Cousins, Emily. "Mountains made alive: Native American relationships with sacred land." CrossCurrents (1996): 497-509.

human worldly expression, and our true humanity emerges through divine encounter.

This spiritual upbringing, coupled with the influences of a strict and conservative family environment, encouraged me to see earthly life as best approached through sacrifice and abstinence, with the real payoff occurring after "*the faithful*" die and go to Heaven. I grew up fearing God's wrath for my inability to follow the spiritual standards set by my parents and our church. Existential fear and anxiety were the most outstanding features of my early spiritual experienced. Meanwhile, God's love, grace, and compassion were given far less attention. The personal world in which I grew up was dominated by a poverty consciousness, causing me to experience a lack of love, acceptance, and positive encouragement. The spiritual attainment bar was too high, causing my sense of self-worth to plummet. I felt unworthy of Christ's love and the love of my parents. Even though my parents did love me, I felt unloved and unlovable.

As I reflect upon my early life, I understand and accept the role I played in my own childhood pain and suffering. From a place of deeper self-understanding, it is possible for us to move beyond blame, shame, and guilt to a place of true self-responsibility and forgiveness. The years have showed me that my struggles were ones that my soul needed to experience to progress spiritually in this lifetime. I have come to see that my early spiritual struggles initially taught me how to find existential freedom, and later they taught me forgiveness, trust, courage, and most importantly unconditional love.

I have come to see that the first eighteen years of my life were blessed with a large extended family that shaped me in many positive ways. Mom and Dad did the best for me that they could. They encouraged me to live a spiritually centered life and to educate myself by going to college. I am thankful for my sister Diana and brother Doug, who shared the same spiritual reality during our younger years. My grandmothers' love grounded me and encouraged me to find my own way in life. My uncles cultivated in me a sense of humor and adventure. Through their stories, my aunts taught me a great deal about my mother and father; things I would not otherwise know or appreciate. In an overall sense, family was the primary stage where I first learned to play many archetypal roles, including the victim, villain, caregiver, conspirator, good shepherd, explorer, rebel, manipulator, philosopher, and jester. I was too young to realize it back then, but life is indeed theater, and a spiritual theater at that.

During my high school years, my intellectual curiosity mushroomed, and I was drawn to other worldviews and philosophies. In my junior and senior high school years, I became deeply interested in existential philosophy because of my desire for personal freedom and self-determination. While my original career goal was to enter the Nazarene Seminary in Kankakee, Illinois after high school graduation to train as a minister, I abandoned that goal after spending the summer of 1969 in Tucson, Arizona. In an act of self-determinism, I enrolled in the fall of 1969 at the University of Arizona to study Anthropology, which offered me an alternative worldview to the conservative Christian theology I was exposed to during my childhood and teen years. The anthropological concepts of evolution, worldview, and cultural relativity helped me see the world in more dynamic and less ethnocentric terms.

While the Nazarene Church, as an active presence, was behind me by 1969, its deep conditioning, rooted in fear, anxiety, blame, and shame, lingered inside me for many years. After considerable self-healing work, this early conditioning eventually was integrated and replaced with a healthier worldview grounded in wholeness and love. My three years in Arizona allowed me to experience life on my own, find new role models, and experiment with new ways of being, relating, and doing.

Since my early days in Tucson, I have explored various spiritual paths, including Native American spirituality, Buddhism, quantum spirituality (a path shaped by consciousness philosophy and quantum physics), and a return to Christianity in the Episcopal and Unitarian Universalist churches. For the past 20 years, the interfaith and interspiritual paths have brought the greatest meaning to me. My recent work with Lakota spirituality has led me to explore new ways to ground my spirituality in everyday living. This is where I am today.

F. My Indian World Experiences and this Book

In Sacred Relationship is an outgrowth of my lifelong interest in Native American culture and spirituality. This interest goes back to my childhood days in the 1950s and 1960s in Belmont County, Ohio, where I spent countless hours in the spring and summer months scouring freshly plowed cornfields for Indian arrowheads. My Grandpap Iannone sparked my first curiosity about Indian artifacts and the humans that made them thousands of years before my time.

Over the years, Grandpap found many arrowheads and other stone tools while tilling his vegetable garden on Rock Hill Pike, just west of Bellaire. Please see the location map in Figure 1 below. The area marked as "*Grandpap's Garden*" is where Grandpap found his arrowheads in the 1940s and 1950s. My grandfather's Indian artifact discoveries, and later my own, set me on a lifelong journey to explore and understand the ancient Native American women and men, who walked the earth in Ohio and elsewhere in the United States. This journey has followed many different paths throughout my life.

American Indian roots in Belmont County and the surrounding Ohio Valley region date back to the Archaic period (ca 10,000 B.P.). By my estimation, most of my grandfather's Indian artifact finds date back to the Late Archaic period (ca 6,000-3,000 B.P.) and throughout the Woodland period (ca 3,500-1,600 B.P). These estimates are supported by my review of Native American artifact records for Belmont County and Eastern Ohio at the Ohio Historical Society.[32]

Figure 1: Rock Hill Area, Belmont County, Ohio

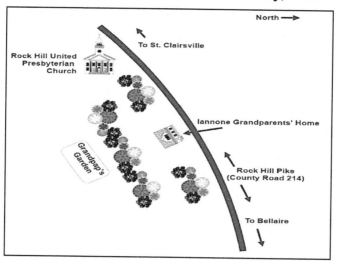

My deep interest in Native American culture led me to select Archeology/Anthropology as my undergraduate major at the University of Arizona (Tucson, Arizona) and Cleveland State University (Cleveland,

---

[32] Source: Ohio Historical Society: https://www.ohiohistory.org/, Accessed December 28, 2019

Ohio). North American Indian Studies was my deepest interest within Anthropology. With the help of two Archeology professors at Cleveland State University, Drs. Douglas McKenzie and John Blank, I gave student leadership to a Middle and Late Woodland Indian village archeological dig along the Cuyahoga River just south of Brecksville, Ohio during the summers of 1973 and 1974. While I was accepted at several major universities with excellent graduate programs in Archeology and Anthropology, I chose to educate myself in other fields of study and chart a career in the economic development and environmental fields.

My American Indian interests were dormant from the mid-1970s until the late 1980s. They were rekindled in the late 1980s while I served as the director of economic development and environmental centers in the College of Urban Affairs at Cleveland State University. In the early 1990s, I applied for and received a multi-year federal grant from the Tribal Environmental Assistance Division of the United States Environmental Protection Agency (USEPA) to provide sustainable development technical assistance and training services to the federally recognized tribes in Michigan, Wisconsin and Minnesota. The work associated with this grant led to many fascinating visits to Indian reservations and tribal lands in these three states.

After leaving Cleveland State University in 2000, I was hired several times to provide economic development and healthcare consulting services to the Cherokee Nation of Oklahoma. This work led to many visits to Tahlequah and other locations in northeastern Oklahoma to work with the Cherokee Nation. My work with the Cherokees deepened my interest in Native American religion.

Since my Cherokee Nation work in the early 2000s, my wife Mary and I have visited many Indian reservations and sacred sites throughout the United States. We have experienced Native American spirituality through vision quests, sweat lodge ceremonies, and medicine wheel walks. During my divinity studies, I conducted extensive research on Lakota culture and spirituality, and I wrote my Master of Divinity thesis on the subject of the relationship between Lakota spirituality and sacred land, which formed much of the research basis for this book. My thesis research included fieldwork on sacred sites in the Black Hills and the Badlands in southwestern South Dakota and Devils Tower in eastern Wyoming.

G. Research Methodology

Research Steps

The research conducted for *In Sacred Relationship* proceeded in six sequential steps:

1. **Literature Review:** A review was conducted of the research literature in several fields. This review examined books, articles, and credible website information about Lakota spirituality, history and culture, Native American sacred lands and sites, and the Black Hills, the Badlands, and Devils Tower. It also included an examination of the literature on grounded and practical spirituality.

2. **Sacred Lands Field Visits:** Field visits were made to the Black Hills and the Badlands nature areas in southwestern South Dakota and Devils Tower in eastern Wyoming, which are considered to be sacred lands by the Lakota people. These areas were explored and experienced from the standpoint of their natural beauty and spiritual significance. These visits *"grounded"* the research that has gone into this book.

3. **Expert Interviews:** Interviews were undertaken with experts on the various aspects of my research topic. These interviews included conversations with a Lakota spiritual teacher, a Lakota artist, and a Lakota tribal elder, as well as interviews with Lakota and North American Indian scholars who have written extensively about Lakota culture, spirituality, and sacred lands. Like the field visits, the expert interviews served to *"ground"* my research.

4. **Develop the Grounded Spirituality and Sacred Relationship Compass (GSSRC):** The next step in the methodology was to develop a grounded spirituality compass or model to guide others in evolving their spiritual beliefs and practices to support and inspire their everyday spiritual lives in practical ways. The development of the model *"grounded"* my research on Lakota spirituality by using it to create a tool that can help others.

5. **Synthesis:** The findings of the literature review, expert interviews, and field visits were analyzed and synthesized. This synthesis integrated the research and experiential components of my research.

6. **Book Writing:** The book was written, incorporating all aspects of my research, as well as some personal observations about my research experience.

Observations about the Research Literature

In my review of the existing research literature, I discovered four major themes:

1. **Credible Literature**: The largest portion of the credible published literature on Lakota spirituality and sacred land is found in Anthropology and North American Indian Studies, and Religion and Spirituality. While a great deal of information on Native American spirituality and religion is available on Internet websites, much of this information is not credible for research purposes.

2. **Traditional versus Contemporary Native American Spirituality**: The largest part of the research literature about Native American spiritual beliefs and practices is focused on *traditional* Native American spirituality and religion. Only limited research has been conducted on *contemporary* Native American religion. Lakota spirituality and religion have been written about extensively by academic scholars and popular writers; much more so than the spiritual traditions of other American Indian tribes. This book develops and offers a more contemporary perspective of Lakota spirituality and sacred land, and we can use its wisdom today.

3. **Sacred Lands and Sites Literature**: A sizeable body of credible research has been conducted on Native American sacred lands and sites. In recent years, this literature has tended to focus on sacred site desecration and protection.

4. **Limited Literature on Grounded Spirituality and Sacred Relationships and the Lakota Contribution:** A great deal has been written about living spiritually in an everyday sense, but very little literature exists on *grounded spirituality* and *sacred relationships* per se, or the applicability of Lakota spirituality to grounded spirituality and sacred relationships. This book aims to help fill this void.

ively

PART II:
LAKOTA SPIRITUALITY
AND SACRED LAND

"Religion is for people who're afraid of going to hell.
Spirituality is for those who've already been there."
— Vine Deloria Jr., Lakota Religious Scholar

In Sacred Relationship

CHAPTER 2: OVERVIEW OF INDIAN TRIBES SHARING GEOGRAPHY WITH SOUTH DAKOTA

"The American Indian is of the soil, whether it be the region of forests, plains, pueblos, or mesas. He fits into the landscape, for the hand that fashioned the continent also fashioned the man for his surroundings. He once grew as naturally as the wild sunflowers; he belongs just as the buffalo belonged."
Luther Standing Bear, Lakota Chief

A. American Indian Tribes in South Dakota

Nine federally recognized Indian tribes *share geography*[33] with South Dakota. Some of these tribes also share geography with other surrounding states, including North Dakota, Nebraska, and Minnesota. South Dakota tribes include the:

1. Cheyenne River Sioux
2. Crow Creek Sioux
3. Flandreau Santee Sioux
4. Lower Brule Sioux
5. Oglala Sioux
6. Rosebud Sioux
7. Sisseton Wahpeton Oyate
8. Standing Rock Sioux
9. Yankton Sioux

Figure 2 below shows the nine Indian tribes sharing geography with South Dakota. The Oglala Lakota Tribe is the largest with 46,000 enrolled members, and the Rosebud Lakota Tribe is the second largest

---

[33] The term "share geography with" is used in recognition of the sovereignty of tribal governments. While tribes may be said to be located in a state, they are "share" geography with states in a legal sense.

with 33,000 enrolled members.[34] Enrolled membership includes those members living on and off the reservation. According to the 2014-2018 American Community Survey, the Native American population living in South Dakota was estimated at 72,263, which is 8.4% of South Dakota's total population. In 2010, 5.2 million people nationally self-identified themselves as Native Americans, which is 1.7 percent of the U.S. population.

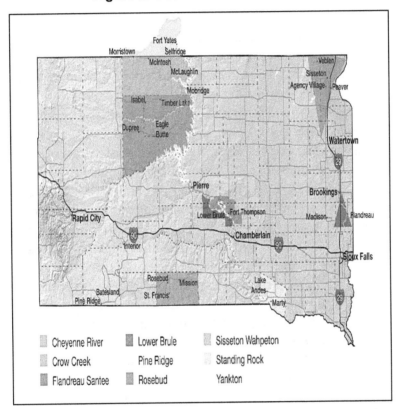

Figure 2: South Dakota Indian Tribes

B. Understanding Tribal Sovereignty and Rights

To fully understand the relationship between American Indian tribes and the U.S. and state governments, it is vital to understand the concept of tribal sovereignty. According to the National Congress of

---

[34] Native Governance Center, South Dakota tribes: https://nativegov.org/south-dakota-tribes/. Accessed December 31, 2019. Note: Enrolled members include those living on and off the reservation.

American Indians: *"Tribal nations are part of the unique American family of governments, nations within a nation, as well as sovereign nations in the global community of nations."*[35] American Indian tribes officially recognized by the U.S. Government are considered sovereign nations for most purposes. There are 573 sovereign Indian Nations within the U.S., which are called tribes, nations, bands, pueblos, communities, and native villages in the United States. In addition, there are state recognized tribes throughout the U.S. that are recognized by their state government.[36]

What is a federally recognized tribe? The U.S. Bureau of Indian Affairs (BIA), says that *"a federally recognized tribe is an American Indian or Alaska Native tribal entity that is recognized as having a government-to-government relationship with the United States, with the responsibilities, powers, limitations, and obligations attached to that designation, and is eligible for funding and services from the Bureau of Indian Affairs. Furthermore, federally recognized tribes are recognized as having certain inherent rights of self-government (i.e., tribal sovereignty) and are entitled to receive certain federal benefits, services, and protections because of their special relationship with the U.S."*[37]

Who is considered to be an American Indian? The U.S. Justice Department says that, *"As a general principle, an Indian is a person who is of some degree Indian blood and is recognized as an Indian by a Tribe and/or the United States. No single federal or tribal criterion establishes a person's identity as an Indian. Government agencies use differing criteria to determine eligibility for programs and services. Tribes also have varying eligibility criteria for membership."* Are American Indians and Alaska Natives citizens of the United States? The Justice Department says they are U.S. citizens, as well as citizens of the states in which they live. They are also citizens of a tribe defined by criteria set by that tribe.

American Indian people believe that the legacy of *"second class citizenship treatment"* by the U.S. Government has grown worse, according to Rep, Raul M. Grijalva (D-AZ.), who said in a 2019 article in The Hill: *"The Trump administration is waging an unprecedented attack on Indian Country. Unless Congress steps up soon, Native Americans across the country could soon lose the ability to determine their own economic*

---

[35] National Congress of American Indians, Tribal Nations and the United States: An Introduction, May 2019.
[36] Ibid.
[37] U.S. Bureau of Indian Affairs: https://www.bia.gov/frequently-asked-questions. Accessed January 1, 2020.

In Sacred Relationship

future."[38] A 2018 U.S. Commission on Civil Rights investigation concluded that: *"Federal programs designed to support the social and economic wellbeing of Native Americans remain chronically underfunded and sometimes inefficiently structured, which leaves many basic needs in the Native American community unmet and contributes to the inequities observed in Native American communities. The federal government has also failed to keep accurate, consistent, and comprehensive records of federal spending on Native American programs, making monitoring of federal spending to meet its trust responsibility difficult. Tribal nations are distinctive sovereigns that have a special government-to-government relationship with the United States. Unequal treatment of tribal governments and lack of full recognition of the sovereign status of tribal governments by state and federal governments, laws, and policies diminish tribal self-determination and negatively impact criminal justice, health, education, housing and economic outcomes for Native Americans."[39]* These conclusions echo those in the Commission's similar 2003 report.

Native Americans remain unconvinced that 2020 U.S. Presidential candidates, especially President Trump, will truly honor their treaty rights, including tribal sovereignty.[40] While tribal nations continue to fight back against legal assaults to their sovereignty, their rights remain in jeopardy under the Trump Administration. Gary Davis, a Forbes Finance Council member said in a recent Forbes Magazine article: *"This is a core principle of how tribes live, operate and govern themselves, but sadly, not everyone respects tribal sovereignty, despite centuries of U.S. Supreme Court and lower court precedent recognizing these powers. Most recently, in a major case for tribal economies, the principle of tribal sovereignty was reaffirmed by the Fourth Circuit Court of Appeals, which demonstrated in its ruling the key role that tribal self-determination plays in economic development throughout Indian Country."[41]*

---

[38] Grijalva, Raul, M., D-AZ, The Trump administration's attack on Indian Country, The Hill, April 11, 2019: https://thehill.com/blogs/congress-blog/energy-environment/438469-the-trump-administrations-attack-on-indian-country. Accessed February 12, 2020.

[39] U.S. Commission on Civil Rights, Broken Promises: Continuing Federal Funding Shortfall for Native Americans, December 2018.

[40] Montgomery, David, What Do Native Americans Want From a President?, The Washington Post, May 13, 2019: https://www.washingtonpost.com/news/magazine/wp/2019/05/13/feature/what-do-native-americans-want-from-a-president/. Accessed February 12, 2020.

[41] Davis, Gary, The Battle Over Tribal Sovereignty: Case Takes Aim At The Use Of Third-Party Vendors, Forbes Magazine, December 12, 2019: https://www.forbes.com/sites/forbesfinancecouncil/2019/12/12/the-battle-over-tribal-sovereignty-case-takes-aim-at-the-use-of-third-party-vendors. Accessed February 12, 2020.

C. Tribal Demographic Overview

American Indian reservations have experienced major social and economic problems from the time they were formed in the mid-19th century. Figure 3 below helps us understand the population size of tribes in South Dakota. These numbers relate to Indians living on South Dakota Indian reservations. Pine Ridge is the largest reservation in terms of population size, followed by Rosebud and Lake Traverse reservations.

**Figure 3: South Dakota Tribal Population Size Comparison
Source: U.S. Census Bureau, Tribal Area Data, 2017**

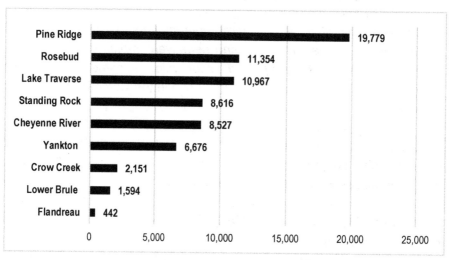

Poverty, unemployment, high educational drop-out rates, poor health, and poor housing conditions are chronic problems among Native Americans living in South Dakota and other states. According to the South Dakota Dashboard, nearly half of South Dakota's Native Americans live in poverty, which is a chronic situation that has remained stagnant since at least 1999, according to U.S. Census data.[42] Many Native Americans and Whites in South Dakota experience high poverty levels, although the level is much higher for Native Americans. For Native Americans, the rate from 2010 through 2014 was 48.4 percent, compared to a 48 percent rate in 1999 and 47 percent from 2005-2009. For Whites, the rate was 9.7 percent

---

[42] My Tribal Area, U.S., Census Bureau: https://www.census.gov/tribal/application/?sec_ak_reference=18.77a43617.1561951312.9644f8embed.html?st=35&aianihh=1700. Accessed September 15, 2019.

In Sacred Relationship

in 1999 and 9.8 percent from 2005-2014.[43] For additional data on South Dakota Indian tribes, the reader is referred to Appendix 1 of this book, which contains a detailed demographic profile of the Pine Ridge Reservation. Appendix 2 provides additional 2017 comparison data for all nine South Dakota tribes.

---

[43] South Dakota Dashboard, Poverty and Income: https://www.southdakotadashboard.org/incomes/poverty, accessed January 1, 2020.

In Sacred Relationship

CHAPTER 3: OVERVIEW OF LAKOTA HISTORY, CULTURE AND SPIRITUALITY

"Crazy Horse dreamed and went into the world where there is nothing but the spirits of all things. That is the real world that is behind this one, and everything we see here is something like a shadow from that one."
Black Elk

A. Lakota History and Culture

The Lakota people live in the American Great Plains Region, primarily in South Dakota. Lakota means *"allies, friends or those who are united."* [44] [45] This book's primary attention is given to the Lakota living in southwestern South Dakota. According to archeological and historical research, the Lakota people have lived in the Black Hills area of South Dakota since the mid-1700s, and since that time their *"homeland"* boundaries have been limited and reset through a series of treaties with the United States Government, wars and conflicts, and other defining events.[46]

Change is a given everywhere. American Indian Country has been far from immune to the influence of external change. The Black Hills area and the larger surrounding Great Plains Region have experienced major changes since the first arrival of the Lakota people. These changes include large inflows of Whites as residents and property owners, the creation of federally designated Indian reservations, where the Lakota and other tribes were forced to live, and the takeover and management of natural

---

[44] "Sioux" Name and Dialects, Akta Lakota Museum and Cultural Center, at: http://aktalakota.stjo.org/site/PageServer?pagename=alm_culture_origins, accessed August 30, 2019.
[45] University of Nebraska, Lincoln, Encyclopedia of the Great Plains at: http://plainshumanities.unl.edu/encyclopedia/, accessed August 31, 2019. The Great Plains is a vast expanse of grasslands stretching from the Rocky Mountains to the Missouri River and from the Rio Grande to the coniferous forests of Canada—an area more than eighteen hundred miles from north to south and more than five hundred miles from east to west. The Great Plains region includes all or parts of Texas, New Mexico, Oklahoma, Kansas, Colorado, Nebraska, Wyoming, South Dakota, North Dakota, Montana, Alberta, Saskatchewan, and Manitoba.
[46] Ostler, Jeffrey. The Lakotas and the Black Hills: The Struggle for Sacred Ground. Penguin, 2010, p. xvii-xvii.

resource areas by the U.S. Government and the State of South Dakota. In addition, significant residential and commercial development have occurred in southwest South Dakota urban areas, including Rapid City, Spearfish, and Sturgis. Tourism, linked to outdoor recreation, scenic beauty, and historical resources, has increased dramatically, especially drawing worldwide visitors to Mount Rushmore, the Crazy Horse Memorial, the Black Hills, and the Badlands. While corporate farming has been prohibited or greatly limited in South Dakota, large-scale agricultural development has grown, forcing many small South Dakota farms out of business.[47]

From a Lakota perspective, many of these development activities have directly and indirectly infringed on Lakota lands, altered traditional Lakota lifestyles, and prompted the acculturation of the Lakota people into the larger and dominant White society.[48] The anthropologist William Powers wrote in 1982, *"The Oglalas at Pine Ridge never quite recuperated from the grim horror of Wounded Knee."* Powers observed that the Lakota people have endured, despite the major

> "The land is sacred. These words are at the core of your being. The land is our mother, the rivers our blood. Take our land away and we die. That is, the Indian in us dies."
>
> ~ Mary Brave Bird, Lakota

challenges to their culture, because of *"Lakȟól wicohan,"* or the *"Indian Way,"* which is a set of beliefs that ranks certain values as superior to those held by *"White Man."*[49] This suggests that over time traditional beliefs and values have played a somewhat stabilizing role in Lakota cultural survival.

B. Two Contrasting Views of Lakota History

According to Mikael Kurkiala, a Swedish anthropologist who has studied Lakota culture, there are two differing accounts of Lakota history. The first is the *"outsider historical explanation,"* which is based in scientific research, and the second is the Lakota oral history and mythological view, which is called the *"insider historical explanation."*[50] An understanding of

---

[47] Corporate Farming Restrictions, South Dakota Legislature:
https://sdlegislature.gov/Statutes/Codified_Laws/DisplayStatute.aspx?Type=Statute&Statute=47-9A. Accessed January 15, 2020.
[48] Deloria, Vine. Red earth, White lies: Native Americans and the myth of scientific fact. Fulcrum Publishing, 1997.
[49] Powers, William, K., Oglala Religion, First Bison Book Printing, November 1982, p. 122-123.
[50] Kurkiala, Mikael. "Objectifying the past: Lakota responses to Western historiography." Critique of anthropology 22.4 (2002): 445-460.

the connection between sacred land and Lakota spirituality requires both explanations. Each is summarized below.

View 1: Western Scientific View of Lakota History

The first view is a version of Lakota history that is widely accepted by a majority of historians and anthropologists, as well as by some Lakota people.[51] It is a history of westward migration and of the Lakota constantly expanding their territories by conquest. The first version says that the Lakota people crossed the Missouri River near Chamberlain, South Dakota around 1760, and that they reached the Black Hills in western South Dakota around 1776.[52][53][54] Many studies conclude that the majority of American Indian tribes were displaced from their original homelands in the eastern, midwestern, and southern regions of the United States during the 18th and 19th centuries.[55][56][57] Siouan language-speaking people[58] are believed to have originated in the lower Mississippi River region, and then later migrated to the upper Ohio Valley region. Pritzker suggests these Siouan people were agriculturalists, and they may have been part of the Mound Builder Civilization during the 9th to 12th centuries CE.[59][60]

As urbanization spread across the eastern and midwestern regions of the United States, conflict over land increased among American Indian tribes during the 19th century.[61] The Lakota are said to have driven the Cheyenne people westward from the Black Hills, and then claimed the territory as their homeland. Later as a tribal confederacy, the *Oceti Sakowin*, which included the Lakota, was able to secure their Black Hills *"homeland"* from the U.S. Government in 1868. This homeland was called

---

[51] Ibid.

[52] Lass, William E. (2000). Minnesota: A History. New York, NY: W. W. Norton & Company.

[53] Schell, Herbert S. (2004). History of South Dakota. Pierre, SD: South Dakota State Historical Society Press.

[54] Kurkiala, Mikael. "Objectifying the past: Lakota responses to Western historiography." Critique of anthropology 22.4 (2002): 445-460.

[55] Zimmerman, Larry J., and Molyneaux, Brian. Native North America, Norman, OK: U. Oklahoma Press, 1996.

[56] Sturtevant, William C. Handbook of North American Indians. Eds. Wilcomb E. Washburn, et al. Vol. 8. Washington, DC: Smithsonian Institution, 1978.

[57] Waldman, Carl, and Braun, Molly. Atlas of the North American Indian. Infobase Publishing, 2009.

[58] Rood, David, S., and Taylor, Allan, R., 1996. Sketch of Lakhota, a Siouan Language. In Ives Goddard (ed.), Languages, 440-482. Washington, D.C.: Smithsonian Institution.

[59] Pritzker, Barry M. A Native American Encyclopedia: History, Culture, and Peoples. Oxford: Oxford University Press, 2000, p. 329.

[60] BCE (Before Common Era) and BC (Before Christ) mean the same thing as before the year 1 CE (Common Era). This is the same as the year AD 1 (Anno Domini), which means "in the year of the lord," often translated as "in the year of our lord."

[61] Bamfroth, Douglas, B., Intertribal Warfare, Encyclopedia of the Great Plains, University of Nebraska Lincoln, at: http://plainshumanities.unl.edu/encyclopedia/doc/egp.war.023, accessed August 31, 2019.

Mni-Sota Makoce, or the *Lakotah Republic*.[62] The map in Figure 4 below shows the changing boundaries of this territory between 1868 and today.

Figure 4: Changing Boundaries of Oceti Sakowin Territory [63]

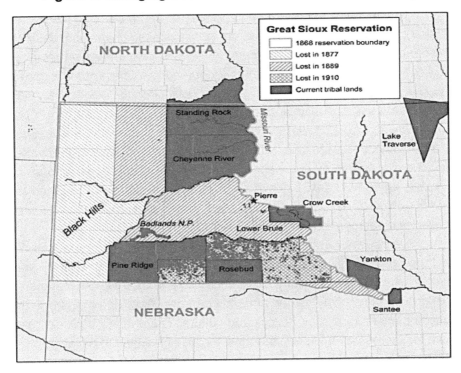

View 2: Lakota History as Seen Rooted in Oral History and Myth

The second view of Lakota history claims that the Black Hills is the birthplace of the Lakota people, which means that the Lakota have lived in the area since ancient times. Kurkiala argues that each version of Lakota history represents a different way of understanding *"truth."* The scientific view is grounded in empiricism and evidenced-based truth. The oral history and mythological view is grounded in experiential understanding. Kurkiala quotes James Walker, an earlier researcher, who described the value of the Native American *"insider"* view: *"In Lakota oral tradition, historical fact was valued not according to its chronological*

---

[62] Treaty of Fort Laramie, 1851, at: https://www.ourdocuments.gov/doc.php?flash=false&doc=42, accessed on August 30, 2019.
[63] Sioux Nation Treaty Council, Great Sioux Reservation at:
https://www.siouxnationtreatycouncil.org/index.php/maps/26-great-sioux-reservation, accessed August 30, 2019.

accuracy, but according to its relevance for the people."[64] Relevance means a great deal in life. It is especially important for Indian people considering their struggles over the past 400 years since European settlers arrived in America.

Kurkiala concludes by saying: *"The need for challenging the White version of history is partly and most significantly based on the land claims, and the Black Hills Claim in particular. If the Lakotas can prove that they were born in the Black Hills and never conquered it, so the argument goes, their case is so much the stronger. Another reason for challenging not only White historians, but also anthropologists, writers, journalists, and film makers say, is more fundamental and less specific. It stems from a feeling of uneasiness in having one's history, identity, spirituality, and problems continuously defined by outsiders. Contesting these outside definitions is, to some, an end in itself. It is a resistance to becoming encompassed in the life world of another; of being made into an object to be understood, studied and categorized."*[65]

I agree with Kurkiala's two-fold explanation of the importance of the *"insider"* version of Lakota history. Lakota history and culture are complex. As within any culture, differences always exist in how members understand their own history and culture, and how non-members or outsiders understand their history and culture. There is still considerable room for improvement in understanding the history of the Lakota and other Native Americans. Both the scientific explanation and cultural worldview offer value in reaching this improved understanding.

Increasingly, Native Americans, including the Lakota, have grown concerned about how their culture, spirituality, and history are understood.[66] Jace DeCory, an American Indian Studies professor at Black Hills State University and an enrolled Cheyenne River Lakota, says it is appropriate for non-Indians to develop their own understanding of Indian culture, history, and spirituality, but they should avoid claims that their views are the same or better than those of American Indians, who should be considered the final authority on these matters.[67]

---

[64] Walker, James (1982) Lakota Society, edited by Raymond J. DeMallie. Lincoln: Univ. of Nebraska Press.
[65] Kurkiala, Mikael. "Objectifying the past: Lakota responses to Western historiography." Critique of anthropology 22.4 (2002): 445-460.
[66] Sanchez, Tony R. "The depiction of Native Americans in recent (1991–2004) secondary American history textbooks: How far have we come?." Equity & Excellence in Education 40.4 (2007): 311-320.
[67] Personal communication Prof. Jace DeCory, Sept. 12, 2019 at Black Hills State University in Spearfish, SD.

C. Advice on Understanding American Indian People

Scholars at the National Museum of the American Indian offer helpful guidelines for understanding American Indians. They remind us of eight important concepts in understanding Native American cultures:[68]

1. There is no single American Indian culture or language.
2. American Indians are both individuals and members of a tribe.
3. American Indians have shaped, and they have been shaped by their culture and living environment.
4. Kinship relationships, including ancestral ties, are essential influences on Indian cultures.
5. American Indian cultures are dynamic and everchanging.
6. Interactions with Europeans and Americans brought accelerated, and often devastating, changes to American Indian cultures.
7. Native people continue to fight to keep the integrity and viability of their indigenous societies. American Indian history is one of cultural persistence, creative adaptation, renewal, and resilience.
8. American Indians share many similarities with other indigenous people of the world, along with many differences.

David Silverman writes in his *Native American Religions: "It is commonplace for scholars of American Indian religions to emphasize the diversity of Indian beliefs and rituals. Nevertheless, Indians across space and time have shared certain religious elements, such as shamanism, the notion of spiritual guardianship, and belief that spiritual power courses through the world and the things in it."*[69] In other words, the similarities across tribes are important and should be recognized.

D. The Oceti Sakowin

The *Oceti Sakowin*, or the Great Sioux Nation, is a Native American confederation that is comprised of three related language families: the Lakota; Dakota; and Nakota. The Lakota is the largest

---

[68] Smithsonian Institutions, Museum of the American Indian, Native Knowledge 360 ° Framework for Essential Understandings about American Indians at: https://americanindian.si.edu/nk360/pdf/NMAI-Essential-Understandings.pdf, accessed September 2, 2019.
[69] Silverman, David J. Native American Religions. Oxford University Press, 2013.

language group, which is made up of seven tribes that live mostly in South Dakota. *Oceti Sakowin* (Och-et-eeshak-oh-win) means Seven Council Fires. In *Oceti Sakowin* culture, the sharing of a common campfire is a symbol of relationship and connection. The Seven Council Fires are the:[70]

- *Mdewakanton* (Dwellers by the Sacred Lake)
- *Wahpekute* (Shooters Among the Leaves)
- *Sisitonwan/Sisseton* (People of the Marsh)
- *Wahpetonwan* (Dwellers Among the Leaves)
- *Ihanktown/Lower Yanktonai* (People of the End)
- *Ihanktowana/Upper Yanktoni* (People of the Little End)
- *Tetonwan* (People on the Plains) (Lakota)

There are seven bands of the *Tetonwan*, or the Teton group. This book concerns itself mostly with the Oglala, who are associated with the Pine Ridge Indian Reservation. The seven Teton bands are the:[71]

- *Hunkpap:* Camps at the Horn (Standing Rock & Wood Mountain)
- *Sicangu/Brule:* (Burnt Thigh Rosebud & Lower Brule)
- *Itazipo/Sans Arc*: Without Bows (Cheyenne River)
- *Sihasapa*: Blackfeet (Cheyenne River & Standing Rock)
- *Oglala:* Scatters His Own (Pine Ridge) (Greatest attention is given to the Oglala Lakota because of their openness in sharing information about their spiritual beliefs and practices.)
- *Oohenumpa*: Two Kettles (Cheyenne River)
- *Mniconjou:* Planters by the River (Cheyenne River)

E. Native American Spirituality Overview

This section provides a general overview of Native American spirituality to set the stage for understanding Lakota spirituality. Anthropologist William K. Powers says that traditional Lakota spirituality, in terms of strict adherence by tribal members, has waned in importance to the Lakota in today's world. In Powers' assessment, the Pine Ridge Reservation has gravitated more to Christianity than traditional Lakota spirituality due to the ongoing influence of the Catholic Church in

---

[70] Oceti Sakowin, Seven Council Fires, Akta Lakota Museum and Cultural Center website: http://aktalakota.stjo.org/site/News2?page=NewsArticle&id=8309, accessed August 30, 2019.
[71] Bonvillain, Nancy. The Teton Sioux. Infobase Publishing, 2005.

particular.[72] Limited research has been conducted on the contemporary religious and spiritual beliefs and practices of Native Americans. The Pew Religious Landscape Survey estimated that less than 0.3 percent of Americans identified themselves with Native American religions.[73]

On the other hand, Native American spirituality is of rising interest to non-Indians. Several observers call attention to the incorporation of Native American spiritual beliefs and practices into New Age spiritual movements, which reflects the growing spiritual eclecticism and diversity found in contemporary Western society.[74] [75] [76] [77] For many, *nature is their church*, which is what draws people to the sweet songs sung by birds on the forest's edge, the sound of the wind whispering its way across the prairie, the sight of a teetering newborn fawn, or the glory of a sunset spread across a lake's glass-like surface.

Spiritual tourism, including Native American spiritual tourism, has grown across the world as spiritual practices such as vision quests, sweat lodge ceremonies, vortex tours, pilgrimages, deep nature experiences, and even drug-induced spiritual experiences have grown in popularity.[78] [79] [80] Millions of visitors flock each year to spiritual tourism destinations such as Macchu Picchu in Peru, Stonehenge in England, the Great Pyramids in Egypt, and The Dead Sea in Israel. In 2018, over 3 million people visited Sedona, Arizona, and another 600,000 people visited Wyoming's Devils Tower, which are major Native American spiritual travel destinations.[81]

Two studies of the religio-spiritual participation of two Native American groups by researcher Eva Marie Garroutte found great diversity in the religious and spiritual participation among Southwestern and Northern Plain Indians. This participation is concentrated in Christianity,

---

[72] Powers, William, K., Oglala Religion, Bison Book Printing, 1982.

[73] Pew Research Center, Religious Landscape Study at: https://www.pewforum.org/religious-landscape-study/#religions, accessed September 25, 2019.

[74] Aldred, Lisa. "Plastic shamans and astroturf sun dances: New Age commercialization of Native American spirituality." American Indian Quarterly 24.3 (2000): 329-352.

[75] Timothy, Dallen J., and Paul J. Conover. "10 Nature religion, self-spirituality and New Age tourism." Tourism, religion and spiritual journeys (2006): 139.

[76] Bowman, Marion. "The noble savage and the global village: Cultural evolution in new age and neo-pagan THOUGHT." Journal of Contemporary Religion 10.2 (1995): 139-149.

[77] Welch, Christina. "Appropriating the didjeridu and the sweat lodge: new age baddies and Indigenous victims?." Journal of Contemporary Religion 17.1 (2002): 21-38.

[78] Stausberg, Michael. Religion and tourism: Crossroads, destinations and encounters. Routledge, 2012.

[79] Sharpley, Richard, and Priya Sundaram. "Tourism: A sacred journey? The case of ashram tourism, India." International journal of tourism research 7.3 (2005): 161-171.

[80] Timothy, Dallen J., and Paul J. Conover. "10 Nature religion, self-spirituality and New Age tourism." Tourism, religion and spiritual journeys (2006): 139-149.

[81] Statistics from Travel and Leisure Magazine, Wyoming Tourism Bureau, Sedona Chamber of Commerce.

aboriginal spirituality (traditional Native American spiritual beliefs and practices), and the Native American Church (The Peyote Way). Belief salience, which is the perceived relevance of spiritual beliefs to everyday life, appears to explain Native American religio-spiritual participation more so than age, gender, or educational factors. Like most other cultural and social groups, members of these two Native American groups show a high degree of *"individualization"* of their religion or spirituality.[82] [83] One-size-fits-all religion has lost favor in the U.S. and other parts of the world. Increasingly, people are seeking more authentic and experiential *"personal relationships with the Divine."*[84] [85] [86] In part, this same trend explains the Charismatic Movement that has grown across Christian denominations and faith communities.[87]

An increasing number of people desire a personal relationship with God, the Divine, or the Holy, which is the experience of the *"mystical interspiritual,"* according to the Catholic monk Wayne Teasdale, author of *The Mystic Heart: Discovering a Universal Spirituality in the World's Religions.* Teasdale says: *"Interspirituality is essentially an agent of a universal mysticism and integral spirituality. We often walk the interspiritual or intermystical path in an intuitive attempt to reach a more complete truth. That final integration, a deep convergence, is an integral spirituality. Bringing together all the great systems of spiritual wisdom, practice, insight, reflection, experience, and science provides a truly integral understanding of spirituality in its practical application in our lives, regardless of our tradition."* [88] The Lakota see themselves in direct relationship to the Great Spirit, other beings, nature, and the universe.

Mysticism has long had its place in Native American spirituality, says Roselyn Marie Amenta, who wrote in her dissertation, *"But the mystical experience indigenous to the native North American tribal peoples stands out uniquely in the history of religions in that it is grounded in the vision of a Sacred World which underlies, envelops and is manifestly*

---

[82] Garroutte, Eva Marie, et al. "Religio-Spiritual Participation in Two American Indian Populations." Journal for the scientific study of religion 53.1 (2014): 17-37.

[83] Garroutte, Eva M., et al. "Religiosity and spiritual engagement in two American Indian populations." Journal for the Scientific Study of Religion 48.3 (2009): 480-500.

[84] Robbins, Jeffrey W. "In search of a non-dogmatic theology." CrossCurrents (2003): 185-199.

[85] Miller, Tracy. US Religious Beliefs and Practices: Diverse and Politically Relevant. 2008.

[86] Wuthnow, Robert. "Creative spirituality: the way of the artist." Nova Religio 9.1 (2005): 123-125.

[87] Setzer, Ed, Understanding the Charismatic Movement, Christianity Today, October 18, 2013 at: https://www.christianitytoday.com/edstetzer/2013/october/charismatic-renewal-movement.html. Accessed September 17, 2019.

[88] Teasdale, Wayne. The mystic heart: Discovering universal spirituality in world religions. New World, 2010.

present within the entire empirical order. The sacred dimension, although being qualitatively different from the phenomenal realm is not regarded as "otherworldly" or "transcendently beyond" in the same way as it is meant in a monotheistic or monistic apprehension of the Divine. In this sense, the Native American does not "transcend" the world in the Christian or Vedantic or Neo-Platonic manner; rather, he "plunges into" the world and experiences himself and the earth with all its myriad phenomena as being totally immersed in and revelatory of the Divine Reality." [89]

Religions and spiritual traditions escape neat definitions because their very essence is *experiential* in nature, making religious experience very individual and unique.[90] [91] Anthropologist David Posthumus helps us understand this phenomenon within Lakota spirituality in saying: *"Lakota tradition, religion, and ritual defy neat categorization and simplistic classifications and demarcations. Tradition is truly an ongoing, organic, and timeless process. Like myth, tradition must be conceptualized as time out of time or atemporal."* [92] Posthumus's observations about the contemporary *"hybridity"* of Lakota religion and spirituality support the research findings by Garroutte summarized earlier.

In my view, the spiritual hybridity found among Lakota people is not unlike the increasingly individualized and diverse spiritual beliefs and practices of people of all religions and cultures. The *"personalization"* of religion and spirituality is essential if it is to have relevance to our lives. This reminds us to go beyond our intellectual understanding and seek our own personal experiential understanding of Lakota spirituality. This reinforces the importance of my fieldwork in Wyoming and South Dakota in researching Lakota sacred lands and sites for *In Sacred Relationship*.

F. Lakota Spiritual Worldview

Lakota Spiritual Beliefs Shared with Other Native Americans

While Lakota spirituality is unique in some respects, it also shares several beliefs and practices in common with other Native American

---

[89] Amenta. Rosalyn, Marie, "The Earth Mysticism of the Native American Tribal People with Special Reference to the Circle Symbol and the Sioux Sun Dance Rite." Doctoral Dissertation. (1987). ETD Collection for Fordham University. AAI8714570. https://fordham.bepress.com/dissertations/AAI8714570

[90] Proudfoot, Wayne. Religious experience. University of California Press, 1985.

[91] James, William. The varieties of religious experience: A study in human nature. Routledge, 2003.

[92] Posthumus, David. "Transmitting Sacred Knowledge: Aspects of Historical and Contemporary Oglala Lakota Belief and Ritual," p. 442-443. PhD dissertation, University of Indiana, 2015.

In Sacred Relationship

spiritual systems. Some of the major shared spiritual beliefs and practices are these:[93] [94] [95] [96]

1. Everything made by the Creator has a spirit. For that reason, all things are related, and all things are sacred. This connected view of life and the universe is an important concept in Native American spirituality.

2. Relationships between humans, Mother Earth, other creatures, and ancestors have an order in the sense that Mother Earth supports all creatures. Animal and plant life sacrifice their lives to nourish and support humans, and they are sacred for this reason. Similarly, ancestors who brought later generations into being are sacred, and they are seen as very much alive in the realm of the spirits, where they guide those living now.

3. Daily life and spiritual life are closely intertwined, and they are not separate. Each has meaning and purpose to the other. For this reason, Native Americans view the earth as sacred because of its role in daily life and because the earth is animated with spirit.

4. Spirit power is seen as existing in all things, which means *"medicine"* is available to support healing of the body, mind, and spirit. Whatever we encounter in daily life is regarded as a *"relative"* to humans. In this sense, everything in life is related by its purpose in serving the Great Spirit or the divine.

5. Some Native American spiritual worldviews see spiritual essence as formless mystical energy or power. This is the case with Wakan Tanka of the Lakota, which is the term used for the Great Spirit as the sacred and divine.

---

[93] Zimmerman, Larry J., and Brian Molyneaux. Native North America. Norman, OK: University of Oklahoma Press, 1996, p. 74-81.
[94] Deloria, Vine. God is red: A native view of religion. Fulcrum Publishing, 2003. P. 77-96 and 113-132.
[95] Smith, Huston. A seat at the table: Huston Smith in conversation with Native Americans on religious freedom. Univ of California Press, 2007, p. 39-57.
[96] McGaa, Ed. Original: Mother Earth Spirituality: Native American Paths To Healing Ourselves And Our World. 1990, p. 113-120.

In Sacred Relationship

6. All Native Americans have rituals and ceremonies to honor Great Spirit, their ancestors, and the land. Native people have their own unique ways of carrying out these sacred purposes.

7. For many Native American people, the concepts of good and evil are defined by whether attitudes and actions do or do not honor human obligations to the spirits.

8. Cosmic balance is a concept shared by many Native people. For this reason, Native American spiritual traditions give considerable attention to the stars and planets and their movement in the skies.

9. Harmony with oneself, other living beings, and the cosmic order is another shared spiritual concept.

10. Nature and spirt are viewed as inseparable and mutually dependent because spirit exists in all things.

11. Native American spiritual systems share the belief in the sacred nature of the circle, and they tend to see life and time in cyclical terms rather than linear terms. Life is often seen as a never-ending cycle of generation (birth), destruction (death), and rebirth.

12. For many Native Americans, the earth is seen as the host for human beings and all other life forms. Humans are generally not seen as any more important than other living or non-living things.

13. Animal spirits are often seen as teachers to humans. For this reason, animals inform human morality and behavior.

14. Land is viewed as sacred because many Native American people see their communities and daily lives as extensions of the spiritually alive natural world.

In spiritual terms, the Lakota people see the world as having four cardinal directions, which is a concept found in several American Indian spiritual traditions. The *medicine wheel* embodies the concept of the four cardinal directions of spiritual significance to the Lakota and other

American Indians. The medicine wheel, or the sacred hoop, has been used by generations of Native Americans as a symbol and tool for health and healing. The medicine wheel is described in this way by the Lakota St. Joseph's Indian School: *"The Medicine Wheel is a sacred symbol used by the indigenous Plains tribes to represent all knowledge of the universe. The Medicine Wheel is a symbol of hope — a movement toward healing for those who seek it."*[97]

In a 1998 article, Richard Roberts discussed the alignment of the Native American medicine wheel with science-based psychology. He reminds us of the limits of scientific psychology in understanding the *"Indian mind."* Roberts urges counseling professionals working with Native Americans to reflect an understanding of how Native Americans view the world and their role and place in it. Roberts helps us understand how cultural worldview and spiritual beliefs are relevant in understanding emotional and psychological healing in the Indian world.[98] The same could be said about how non-Indians seek to understand Indian sacred land.

Within traditional Lakota spirituality, each of the four cardinal directions has a special meaning and color. Here is a summary of the four directions and their significance:[99] [100]

1. **East (Yellow) Wioheumpata**: This is the direction from the sun, which signifies the morning light, the beginning of a new day, and the beginning of all spiritual understanding. Light helps us see reality. The East stands for the wisdom that people need to live good lives that are in balance personally, socially, and cosmically.

2. **South (White) Itokaga**: The sun is at its highest point in the southern sky, which signifies warmth and growth. The Lakota believe that the life of all things originates in the south, and after death, people pass into the spirit world by way of the Milky Way's path back to the south, which is where spirits originate.

---

[97] The Medicine Wheel, St. Joseph's Indian School website: https://www.stjo.org/native-american-culture/native-american-beliefs/medicine-wheel/, accessed September 28, 2019.
[98] Roberts, Richard L., et al. "The Native American medicine wheel and individual psychology: Common themes." Journal of Individual Psychology 54 (1998): 135-145.
[99] Akta Lakota Museum and Cultural Center, The Four Directions, at: http://aktalakota.stjo.org/site/News2?page=NewsArticle&id=8593, accessed September 4, 2019
[100] Goodman, Ronald, Lakota Star Knowledge; College Park, Md. Vol. 12, (Jan 1, 1996): 140.

In Sacred Relationship

3. **West (Black) Wiyokpiyata**: The sun sets in the west and each day ends. The Lakota see the west as signifying life's end. The west is also the source of all water, including the rain, lakes, streams, and rivers. Water is a Lakota life essential.

4. **North (Red) Waziyata**: The cold of winter comes from the north. While harsh, the winds are cleansing. Facing the harsh winds is a test of courage and strength. For that reason, the winds of winter teach us patience and endurance.

The *"four directions concept"* helps us understand the intersection of everyday life and spiritual reality in Lakota culture. Nature and Mother Earth's powers are apparent in each of the four directions. In his book, *Mother Earth Spirituality*, Ed McGaa (Eagle Man) describes the important connection between nature and Lakota spirituality in this way: *"Native American Indians learned how to live with the earth on a deeply spiritual plane. Their intuitive sense of intimate connection with all of existence from Brother Bear to Sister Stone to Father Sky to Mother Earth provides the deep ecological wisdom that the present-day environmental prophets have rediscovered and begun to teach to an alienated world."*[101] McGaa reminds us that ecological balance and spirituality go hand in hand in the Lakota spiritual worldview. Tribal environmental policies are often based upon spiritual worldviews, while U.S. and state environmental policies lack spiritual grounding.[102]

Unique Aspects of Lakota Spirituality

Lakota spirituality is not easily understood as a religion, which is a term that grows out of Western ideology and Christianity. According to the Encyclopedia of Religion, *"For the Lakota, religion is not compartmentalized into a separate category. More appropriately, Lakota traditions can be characterized as a system of spirituality that is fully integrated into a rhythm of life that includes all aspects and patterns of the universe. At the center of this rhythm is Wakan Tanka or Tunkashila, sometimes translated as Grandfather and often as Great Spirit or Great*

---

[101] McGaa, Ed., Mother Earth Spirituality: Native American Paths to Healing Ourselves and World. 1990, p.xiv.
[102] Warner, Elizabeth Ann Kronk. "Looking to the third sovereign: Tribal environmental ethics as an alternative paradigm." Pace Envtl. L. Rev. 33 (2015): 397.

Mystery, but better left untranslated."[103]

Religion can be seen as a belief system, rooted in culture, which provides a model of reality, according to William K. Powers, an anthropologist specializing in Oglala Lakota culture.[104] Powers encourages us to see Oglala Lakota spirituality in four contexts: 1) within its larger sociopolitical context, which is in relationship to the Oceti Sakowin (the Seven Council Fires); 2) within the context of Lakota *"sacred things,"* especially how the Lakota view supernatural beings and powers; 3) within the Lakota concept of *"mitakuye oyasin,"* or the connectedness of all things, or as the Lakota say, *"all my relations;"* and 4) how the Lakota explain reality in synchronic terms (at a particular point in time without reference to historical antecedents), and in diachronic terms (at various points across or through history).[105]

Building on the ideas of earlier scholars, Powers closes his book with this reminder, *"Man has essentially two conflicts to resolve: his rather indeterminate position between nature and culture, and his need to reconcile the irreversibility of death."*[106] Regarding the first conflict, Powers attributes great significance to social relations, kinship, and ancestor relations. The ordering of the Lakota universe reflects the view that cultural things are analogous to natural things, which helps explain the role played by sacred lands and how the Oglala Lakota live in everyday life.[107] With respect to the irreversibility of death, Powers says, *"The need to reconcile the irreversibility of death is overcome in the reincarnative system of the Oglalas. The distinction between life and death in such a system is blurred. The ghosts of the dead appear at will and communicate with the living. The souls of the living disappear during rituals and communicate with the deceased. The two worlds are mirror images of each other, and they are conjoined by a continuum, an ethereal road which passes across the sky in a southerly route and dips beneath the earth, only to reemerge again in the north, the land of pine, and breath, and life."*[108]

Within the Lakota worldview, space determines the *nature of things*, which explains how sacred land is understood, and it accounts for

---

[103] Jones, Lindsay, and Mircea Eliade. The encyclopedia of religion. Macmillan Reference USA, 2005, Lakota Religious Traditions, p. 5295-5298.
[104] Powers, William, K., Oglala Religion, First Bison Book Printing, November 1982, p. xv.
[105] Ibid, p. xx-xxi.
[106] Ibid, p. 188-189.
[107] Ibid, p. 188-191.
[108] Ibid, p. 200-201.

the importance of space (specific locations and directionality) in Lakota ceremonies and settlement patterns.[109] Time, on the other hand, determines the *meaning of relationships*, which is illustrated by: 1) the four times (day, night, month, year); 2) the seven generations; 3) the larger cycles of time connecting specific locations, including sacred places, to the larger universe; and 4) Lakota star knowledge, which is seen as helping the Lakota people to understand the temporal-spatial dimension and how everyday life events were timed to mirror celestial movements.[110]

David Posthumus, a cultural anthropologist at the University of South Dakota who has studied the Oglala Lakota spiritual tradition extensively, describes how Oglala Lakota spirituality has become widely accessible to Indians and non-Indians, *"The Oglalas are especially renowned for their poignant religious philosophies and eloquent religious leaders and thinkers. Brilliant and exceptional individuals such as George Sword, Horn Chips, Nicholas Black Elk, Frank Good Lance, Frank Fools Crow, and Peter Catches, Sr. found ingenious ways, despite settler colonialism and difficult historical circumstances, to translate their spiritual beliefs and traditions in such a way as to make them accessible to all."*[111] Posthumus goes on to point out that Oglala Lakota spirituality has persisted beyond that of many Native American spiritual traditions, which in his view is because *"the Oglalas do these things purely out of a deep love and respect for Lakȟól wičȟóȟ'aŋ or Lakota traditions and the Lakota way of life."*[112] Posthumus says that while the old religious leaders such as Black Elk, Horn Chips, and Little Warrior, are long gone, new leaders have stepped up to keep the Oglala Lakota spiritual tradition alive, while renewing it with new visions of the Lakota people in the future.

Zimmerman and Molyneaux remind us that Native American spiritual traditions are *"localized"* to incorporate the everyday life of Native people.[113] Lakota spirituality in western South Dakota and eastern Wyoming is very much shaped by the plains, the mountains, and forests

---

[109] Black Elk, N. (1980). Wiwanyag Wachipi: The sun dance. In J. E. Brown (Ed.), The sacred pipe: Black Elk's account of the seven rites of the Oglala Sioux (pp. 67-100). Norman: University of Oklahoma Press.
[110] Ibid, p. 3-9.
[111] Posthumus, David. "Transmitting Sacred Knowledge: Aspects of Historical and Contemporary Oglala Lakota Belief and Ritual," p. 1-2, PhD dissertation, University of Indiana, 2015.
[112] Ibid, p.3.
[113] Zimmerman, Larry J., and Brian Molyneaux. Native North America. University of Okla. Press, 1996, p. 74-81.

In Sacred Relationship

found on and surrounding their reservations. Posthumus[114] and Osler[115] make this same point. Black Elk's vision remains central to the Lakotas' insistence that their very identity hinges on their free presence in the Black Hills. Osler writes, *"Toward the end of Black Elk's vision, a spirit called on him and said, 'Take courage, for we shall take you to the center of the earth.' Black Elk looked and saw great mountains with rocks and earth."*[116] Black Elk was talking about the Black Hills, which is sacred land of great importance to the Lakota. Lakota spirituality is inseparable from the local Black Hills landscape. Specific places, like the Black Hills, Badlands, and Devils Tower, give unique meaning and power to Lakota spirituality.

In addition to the power of specific places, the myths and oral traditions of the Lakota people give uniqueness to Lakota spirituality. According to Lakota myth, before creation, Wakan Tanka (Great Spirit) existed in a great emptiness called Han (or in the darkness). Because of his need for companionship, Wakan Tanka created other spiritual entities. First, he used his creative energy to form Inyan (Rock), which was the first Lakota god. Secondly, Wakan Tanka used Inyan to create Maka (earth), and then he mated with that Maka to produce Skan (sky). Skan brought forth Wi (the sun) from Inyan, Maka, and Wakan Tanka. While all four of these gods are separate and powerful in their own right, each remains a permanent part of Wakan Tanka. [117]

A final unique feature of Lakota spirituality is the teachings of Lakota religious leaders, such as Black Elk and Crazy Horse, who set *"living examples"* of Lakota spiritual beliefs and how spiritual rituals (sweat lodge ceremonies, vision quests, and pipe rituals) create a living understanding of Lakota spirituality.

---

[114] Posthumus, David. "Transmitting Sacred Knowledge: Aspects of Historical and Contemporary Oglala Lakota Belief and Ritual." (2015), p. 1-2, PhD dissertation, University of Indiana, p. 56-58.
[115] Ostler, Jeffrey. The Lakotas and the Black Hills: The Struggle for Sacred Ground. Penguin, 2010, p. 3-7.
[116] Ibid, p. 77.
[117] Walker, James R. Lakota myth. U of Nebraska Press, 2006, p. 10-12, and Erdoes, Richard, and Alfonso Ortiz, eds. American Indian myths and legends. Pantheon, 1984, p. 50-51.

CHAPTER 4: UNDERSTANDING SACRED LAND

"This mountain has my heart. This land is our church."
Caleen Sisk, Winnemem Wintu Tribe

A. Sacred Land Definitions

Land is considered to be *"sacred,"* or spiritually significant, by the Lakota, many other Native American tribes, and indigenous cultures across the world. The worldview of many indigenous cultures includes the principle of *spiritual connectivity*, which says that everything in the universe is connected by spirit. *Spiritual connectivity* explains how land and human spirit are related in a spiritual sense. As mentioned earlier, the Lakota call this *"mitakuye oyasin,"* or *"all my relations."* We exist in relationship to all beings and all things.

Peter Nabokov described the American Indian's intimate experience with the land in this almost poetic way, *"They cherished the places where they gathered ritual materials, the meadows where they collected plants, the rapids and riverbeds where they fished, and the woods, seas, and plains where they hunted. They reciprocated with offerings to the plants they harvested and the animals they killed. All over North America the landscape is saturated with Indian memories and stories that describe such beliefs. Of course, these practices differed from habitat to habitat, but is fair to say that Indians played a part in the inner life of land, and it responded as an influential participant in theirs."*[118]

Lisa A. McLoughlin says that land is central to spiritual meaning in earth-based cultures, including American Indian cultures. McLoughlin quotes Vine Deloria, a prominent Indigenous American scholar: *"My culture deems the land as its own sentient, religious force. Religion cannot be kept within the bounds of sermons and scriptures. It is a force in and of itself, and it calls for the integration of lands and peoples in harmonious*

---

[118] Nabokov, Peter. Where the lightning strikes: The lives of American Indian sacred places. Penguin, 2007, p.5.

unity. The lands wait for those who can discern their rhythms." [119] [120]

How do sacred spaces and places differ from other types of spaces and places? Diane Osbon draws upon the comparative mythologist Joseph Campbell's views of sacred space and place in her writing: *"A sacred space is any space that is set apart from the usual context of life. In the secular context, one is concerned with pairs of opposites: cause and effect, gain and loss, and so on. Sacred space has no function in the way of earning a living or a reputation. Practical use is not the dominant feature of anything in the space."*[121]

Jill Oakes from the University of Manitoba defines sacred land in the Indian world in this way: *"Sacred lands are spaces that hold special significance or values or meaning to specific groups and could be as small as a ring, invisible as a mental space, or as large as Mother Earth and the Universe."*[122] This is consistent with the discussion above about Native American spirituality in general and Lakota spirituality in specific. Many Native American spiritual traditions view the entire universe as sacred, and they believe that land is imbued with the universe's spiritual powers.

B. How Land is Given Sacred Meaning

How does land become sacred? This is an important question explored in the book *American Sacred Space* by Chidester and Linenthal, which says there are three basic views of how land or space become known as sacred. The first is through the *identification with special spiritual powers or energies inherent in certain places or lands*, which cause them to be seen as sacred in nature. The second approach is by *"sacralization"* through cultural ritual. This reminds us of the importance of the reenactment of creation stories through spiritual ceremonies. The third way combines the *"inherent spiritual power present within the place"* approach and the *"sacralization through ritual and ceremony"* approach. In the third approach, both ceremony and inherent spiritual presence are seen as necessary for the sacred to exist and be experienced.[123]

---

[119] McLoughlin, Lisa A. "US Pagans and Indigenous Americans: Land and Identity." Religions 10.3 (2019): 152.
[120] Deloria, Vine. God is red: A native view of religion. Fulcrum Publishing, 2003.
[121] Osbon, D. K. (1991). Reflections on the Art of Living. A Joseph Campbell Companion. United States of America: Harper Perennial, p. 180.
[122] Oakes, Jill. "Sacred Lands: Aboriginal World Views." Claims and Conflicts, Edmonton, Alberta: Canadian Circumpolar Institute, University of Alberta (1988).
[123] Chidester, David, and Edward Tabor Linenthal, eds. American sacred space. Indiana University Press, 1995, p. 5-7.

In Sacred Relationship

The *experience of the sacred* is important to understanding the experience of land as sacred. The German theologian and philosopher Rudolph Otto uses the term *"numinous"* in his book, *The Idea of the Holy*, to describe how humans experience the *"Holy"* or the *"Divine."*[124] Numinous derives from the Latin word *"numen,"* which means to arouse spiritual or religious emotions, or a sense of the mysterious or awe-inspiring. The term is also used to refer to the *"deity or spirit presiding over (or residing within) a thing or space."* The *numinous* is an explanation of how land is experienced as sacred by the Lakota and others.

Scientific research in the past two decades suggests that human beings are *"wired for God,"* which is the subject of Charles Foster's book *Wired for God, the Biology of Spiritual Experience.*[125] In this sense, the Lakota, like all of us, are *"wired for God."* Foster's research says the neurology of religious experience shows that during worship and prayer, a part of the brain, apparently dormant during other activities, becomes active.[126] René J. Muller wrote in a 2008 *Psychiatric Times* article: *"Neurotheologians argue that the structure and function of the human brain predispose us to believe in God. They claim that the site of God's biological substrate is the limbic system deep within the brain, which has long been considered to be the biological center for emotion. Rhawn Joseph, a prominent neurotheologian, goes a step further to suggest that the limbic system is dotted with "God neurons" and "God neurotransmitters."*[127] [128] These scientific theories do not in any way *explain away* the presence *of spirit* at work in both humans and things, including the land itself. Instead, this research suggests that the human body is equipped in neurological and physical terms to experience the divine, holy, and numinous.

David Wishart says in the Introduction to the *Encyclopedia of the Great Plains Indians* that the Indian presence itself brings sacredness to the land. Wishart says: *"Finally, it cannot be emphasized too much that over time Indians have endowed the physical environments of the Plains with great sacred significance. This is a consecration of environment and*

---

[124] Otto, Rudolph. "The idea of the holy. (Harvey, J., Trans.) New York." (1958).

[125] Foster, Charles. Wired For God?: The biology of spiritual experience. Hachette UK, 2011.

[126] Ibid.

[127] Muller, Rene, J., Neurotheology: Are We Hardwired for God?, Psychiatric Times, May 2, 2008, accessed on January 8, 2020 at https://www.psychiatrictimes.com/neurotheology-are-we-hardwired-god

[128] Joseph R. The limbic system and the soul: evolution and the neuroanatomy of religious experience. Zygon. 2001; 36:105-136.

place that is generally missing among other populations of the region. The Indians' traditional religions were place-based, the very names of the months in their native languages chronicling changes in the local environment throughout the years."[129]

According to Geoffrey Simmins at the University of Calgary, *"Natural areas intersect the physical and metaphysical worlds. This is particularly true of mountains, caves, and springs (and some groves), rocks, canyons, and of mountains. A notable example would be Acamo Pueblo in the southwest United States: this is not just the oldest continually inhabited location in North America, it is regarded as inhabited because of its sacredness. Certain places are regarded as places where the spirit may be transformed. But there are many kinds of transformation."*[130] Examples of "transformational places" are places for:

- Life passages (puberty, marriage, death)
- Vision questing/ pilgrimage (revelation of power)
- Sweat lodges (imbue participants with special power)
- Sun dances, rain dances, and other ritual activities (for developing and enhancing community, spirit, and for healing)

Legal definitions of sacred land exist to use the power of law to protect sacred lands and protect Native Americans' right to freedom of religion. According to a 1996 Presidential Executive Order, *"Sacred site means any specific, discrete, narrowly delineated location on Federal land that is identified by an Indian tribe, or Indian individual determined to be an appropriately authoritative representative of an Indian religion, as sacred by virtue of its established religious significance to, or ceremonial use by, an Indian religion; provided that the tribe or appropriately authoritative representative of an Indian religion has informed the agency of the existence of such a site."*[131]

The map in Figure 5 below provides a regional view of Devils Tower in eastern Wyoming and the Black Hills and the Badlands in

---

[129] Wishart, David, J., Editor, Encyclopedia of the Great Plains Indians, University of Nebraska Press, 2004, p. vii-viii.

[130] Simmins, Geoffrey, Sacred Spaces and Sacred Places, Book draft accessed at:
https://dspace.ucalgary.ca/bitstream/handle/1880/46834/Sacred%20Spaces.pdf;jsessionid=B2B030886651BE3 1E5E5349214B504F6?sequence=1 on September 4, 2019

[131] Clinton, William, J., President, Indian Sacred Sites Executive Order, May 24, 1996 at:
https://www.nps.gov/history/local-law/eo13007.htm, accessed September 2, 2019.

western South Dakota. Devils Tower is located near Hulett, Wyoming. Hill City, South Dakota is the approximate center of the Black Hills. Wall, South Dakota is the northern entrance to the Badlands. Hulett is located 109 miles northwest of Hill City. Wall is located 80 miles northeast of Hill City. Hulett is located 145 miles northwest of Wall.

Figure 5: Devils Tower-Black Hills-Badlands Area Map[132]

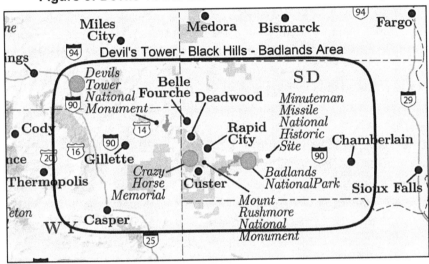

C. Some Key Issues Related to Sacred Lands

A variety of complex issues are associated with sacred lands and sites; not the least of which is the definition and meaning of lands and sites that are considered to be sacred. According to the Sacred Lands Film Project, *"To the Western ear, 'sacred' may be synonymous with 'sacrosanct' — inviolably holy — but to an indigenous culture, a place labeled as 'sacred' may instead mean something spiritually alive, culturally essential, or simply deserving of respect."*[133]

Vine Deloria, a leading Lakota religious scholar, discusses his personal experience in coming to an understanding of the meaning of the sacredness of place: *"I came to revere certain locations and passed the stories along as best I could, although visits to these places were few and*

---

[132] Original map from Visit South Dakota at: https://www.travelsouthdakota.com/ (Annotations added by this author), September 18, 2019.
[133] Sacred Land Film Project, What is a Sacred Site? Website: https://sacredland.org/tools-for-action/. Accessed September 4, 2019.

far between. It seemed to me that the remembrance of human activities at certain locations vested them with a kind of sacredness. Gradually I began to understand a distinction in the sacredness of places. Some sites were sacred in themselves, others had been cherished by generations of people and were now part of their history and, as such, revered by them and part of their very being."[134]

Threats to Indian Sacred Lands

For Native Americans, the centuries-old struggle to retain ownership and control of their homelands is a major issue for sacred lands and sites.[135] [136] [137] This struggle is far from over, as new threats have emerged around the Dakota pipeline,[138] [139] and other major developments. Kyle White, a philosopher at Michigan State University, calls the Dakota pipeline an act of U.S. colonialism, not unlike that seen in early U.S. history and intrusive U.S. Government actions abroad.[140] Walter Mengden IV says this about the Dakota pipeline situation and broader relations between tribes and the U.S. Government: *"Even though tribes are partially sovereign entities, the antiquated idea of a guardian protecting its ward still permeates throughout U.S. government policy. There needs to be a more democratic relationship between Native American tribes and the government. This idea of tribes as domestic dependent nations, that tribes are so alien that they do not have a more defined place in the eyes of the U.S. government, are beliefs of the past."*[141] The Dakota Pipeline is one of many development projects seen by Native Americans as a threat to their way of life, cultural resources, and natural environment. This history is long and complex, as Ashley Glick says in her article.[142] Glick provides a balanced assessment of the history of legal disputes between Native

---

[134] Deloria, Vine. God is red: A native view of religion. Fulcrum Publishing, 2003, p. xv.

[135] LaDuke, Winona. All our relations: Native struggles for land and life. Haymarket Books, 2017.

[136] Churchill, Ward. "The black hills are not for sale: A summary of the Lakota struggle for the 1868 treaty territory." The Journal of Ethnic Studies 18.1 (1990): 127.

[137] Ostler, Jeffrey. The Lakotas and the Black Hills: The Struggle for Sacred Ground. Penguin, 2010.

[138] Whyte, Kyle. "The Dakota access pipeline, environmental injustice, and US colonialism." Red Ink: An International Journal of Indigenous Literature, Arts, & Humanities 19.1 (2017).

[139] Diotalevi, Robert Nicholas, and Susan Burhoe. "Native American Lands and the Keystone Pipeline Expansion: A Legal Analysis." Indigenous Policy Journal 27.3 (2017).

[140] Whyte, Kyle Powys, The Dakota Access Pipeline, Environmental Injustice, and U.S. Colonialism (February 28, 2017). Red Ink: An International Journal of Indigenous Literature, Arts, & Humanities, Issue 19.1, Spr, 2017.

[141] Mengden IV, Walter H. "Indigenous People, Human Rights, and Consultation: The Dakota Access Pipeline." American Indian Law Review 41.2 (2017): 441-466.

[142] Glick, Ashley A. "The Wild West Re-Lived: Oil Pipelines Threaten Native American Tribal Lands." Vill. Envtl. LJ 30 (2019): 105.

Americans, private corporations, and the U.S. Government.[143]

According to the pipeline development company Energy Transfer LP, *"The Dakota Access Pipeline is a 1,172-mile underground 30" pipeline transporting light sweet crude oil from the Bakken/Three Forks production area in North Dakota to Patoka, Illinois."*[144] It says that the pipeline has been *"safely operating since June of 2017, and the Dakota Access Pipeline now transports 570,000 barrels of oil per day."*[145] Despite these claims and the economic benefits cited by the pipeline owner, Native Americans remain highly concerned about the presence of the Pipeline.

Spirituality and the land are inseparable in the eyes of many Native Americans. Joel Brady argues that the U.S. Government should give much greater recognition to the sacredness of land in negotiating land ownership rights with American Indian people. Brady says: *"At the center of most Native American belief systems is the basic tenet that religion and faith draw heavily upon sacred lands. Land is, as Justice Brennan notes, a living being. As such, courts would do well to heed Justice Brennan's delineation of the paramount importance of sacred land to every aspect of Native American life, not just religion. The pervasiveness of spirituality has fueled many Native Americans to be increasingly assertive in exercising their sovereignty. Specifically, they are 'demanding that agencies such as the Park Service treat them like living cultures, not dead ones.' Once the federal government grasps the idea that Native American land is part of the rich tapestry that binds tribal members together as well as an actual, living being in the minds of Native Americans. The relief sought by sacred site protection advocates may finally be realized."*[146]

Some Personal Thoughts on Balancing Future Priorities

The lives and well-being of American Indian people and their sacred lands are complexly intertwined with the whole of American society, which complicates the challenge of creating and preserving boundaries between what is considered sacred in the Indian world and the larger flow of life. I would argue that both non-Indians and Indians need to cultivate greater awareness, sensitivity, and respect for the needs of the planet and our roles as stewards of the planet's future well-being. We

---

[143] Ibid.
[144] Dakota Access Pipeline website: https://www.daplpipelinefacts.com/. Accessed January 8, 2020.
[145] Ibid
[146] Brady, Joel, "Land Is Itself a Sacred, Living Being": Native American Sacred Site Protection on Federal Public Lands Amidst the Shadows of Bear Lodge , 24 American Indian Law Review 153 - 185 (2000)

should treat the earth as sacred in our own ways of knowing and believing. Sustainability is a shared responsibility for everyone. In the case of sacred land in Indian world, the U.S. Government and non-Indians should recognize that these lands hold special meaning and significance to American Indian people, and we should work harder to respect American Indians' right to live in harmony with their sacred lands.

In today's world, we want it all. This is true inside and outside American Indian Country. We want to increase wealth creation and economic growth at almost any economic and non-economic cost. At the same time, we want to live long and healthy lives, we want high quality natural resources, we want to honor religious freedom, and we want to protect cultural resources and spiritually significant lands and sites. We cannot have it all! Something must give.

As I look across American history since the early 1600s, a longstanding pattern of conquest and conflict has prevailed. This pattern has favored wealthy and powerful private interests over Native American interests and the interests of most American people. We have imperiled our natural environment both with and without legal approval. We have selectively told the story of our nation's history to justify our favored attitudes and actions. The American political economy, while idealized across the globe for many years, has come into increasing question by many Americans and people across the world on whether our economic and political ends justify the means we have used to achieve these ends. Lakota spirituality and its reverence for the land calls the American moral compass into question, and how that compass has justified our long history of conquest and conflict favoring the few over the many.

CHAPTER 5: LAKOTA SACRED LAND: BLACK HILLS, BADLANDS, DEVILS TOWER

"The Holy Land is everywhere"
Black Elk

THE BLACK HILLS AS SACRED LAND

A. Black Hills Overview

This section describes the Black Hills geologic history, the Lakota cultural history in the Black Hills, and the Lakota spiritual connection to the Black Hills. Black Hills National Park, which includes much of the area considered sacred by the Lakota people, is 65 miles wide and 125 miles long. The area includes 3,000 acres of wilderness. The Black Hills have been hunting grounds, as well as sacred land, for the Lakota for almost 250 years. The Lakota have lived in harmony with the Hills throughout this history.

The conflict over control of the Blacks Hills sparked the Black Hills War in 1876, also known as the Great Sioux War, which was the last major Indian War on the Great Plains. Upon defeat of the Lakota and their Cheyenne and Arapaho allies, the U.S. Government took back control of the Black Hills. In 1980, in the legal case *United States v. Sioux Nation of Indians*, the U.S. Supreme Court ruled that the Black Hills were illegally taken by the federal government. The court ordered remuneration of the initial offering price plus interest, amounting to nearly $106 million. The Lakota refused the settlement because they wanted the Black Hills returned to tribe. The award money is in an interest-bearing account, which amounted to more than $1.2 billion in 2011.[147] The deepest desire of the Lakota people is to regain ownership of the Black Hills.

---

[147] Streshinsky, Maria (February 9, 2011). "Saying No to $1 Billion". The Atlantic. Accessed January 15, 2020.

In Sacred Relationship

B. Black Hills Geological History[148]

Four primary geological formations are found in the Black Hills in western South Dakota and eastern Wyoming:

1. **Precambrian Formations**: South Dakota's oldest rocks were formed in the Precambrian Period, which was more than 2 billion years ago. These rocks consist of the granites and metamorphic found in the core of the Black Hills. By the end of the Precambrian Period, or about 570 million years ago, South Dakota was deeply eroded and worn to a nearly flat plain interrupted by low knobs of granite and ridges of quartzite.

2. **Paleozoic Formations**: During the early Paleozoic Period (541 to 252 million years ago), the first sea developed and covered western South Dakota. For most of the Paleozoic era, water lapped on and off the state repeatedly. With each advance and retreat, new rock layers of sandstone, shale, and limestone were deposited.

3. **Mesozoic Formations**: The beginning of the Triassic period (250 million years ago) in South Dakota was a time of erosion. Late in the Triassic period, shallow continental seas submerged parts of western South Dakota. During Cretaceous times, the birth of the Black Hills occurred. As the Black Hills area domed upwards, erosion broke down the soft shales, and finally the harder and older sediments.

4. **Tertiary Formations**: Erosion continued as the Tertiary Period began. By the Early Oligocene time (65 million years ago), streams could no longer carry away their erosion products, and earth and rock deposit occurred west of the Black Hills in what is now known as the Badlands.

The Black Hills have been described by author Laural Bidwell in the following way: *"The Black Hills of South Dakota are frequently referred to*

---

[148] Black Hills and Badlands Geology: https://blackhillsvisitor.com/learn/black-hills-and-badlands-geology/2/, accessed September 6, 2019.

as an 'Island in the Plains,' and the description is an apt one. The Cheyenne River sets the southern border while the Belle Fouche River (pronounced Bell Foosh) defines the northern edge. The Thunder Basin National Grassland of Wyoming to the west and the Buffalo Gap National Grassland to the east complete the circle around the hills. The hills rise over 3,000 feet above the plains, reaching their pinnacle of 7,242 feet at Harney Peak, one of the highest points in North America east of the Rocky Mountains."[149]

Photograph 1: Black Hills Forest in Hot Springs, South Dakota[150]

C. Lakota Cultural Life and History in the Black Hills

Introduction

As discussed earlier, the Lakota are one of the Seven Council Fires (*Oceti Sakowin*), also known as the Great Sioux Nation. The *Oceti Sakowin* consists of four Dakota bands (Mdewakanton, Wahpekute, Wahpeton, Sisseton), two Nakota bands (Yankton, Yanktonai) and a Lakota band (Teton). The Lakota culture, like that of most American Indian

---
[149] Bidwell, Laural, A., Geography and Geology of the Black Hills: https://www.moon.com/travel/trip-ideas/geography-geology-black-hills/, accessed September 6, 2019.
[150] Photograph by Don Iannone in Hot Springs, South Dakota on September 10, 2019.

tribes, is a long history of the struggle to survive.

The Black Hills is viewed by the Lakota people as central to Lakota survival because their culture is interlocked with the Hills. Benjamin Jewell, a cultural anthropologist at Arizona State University wrote in a 2006 article: *"The thrust of the U.S. Government's attempts to 'civilize' and save the Lakota was enacted in complete ignorance to cultural values and ideals. Having existed as a nomadic hunting and foraging people for the previous 200 years, if not more, the Lakota were not going to shake their culture's approved way of life readily. Part of the belief system of the Lakota is a spiritual connection to the earth and their homeland, the Black Hills. According to Standing Bear, 'Of all our domain, we loved, perhaps, the Black Hills the most.'*[151] The Lakota named these hills *He Sapa*, or Black Hills, because of their color. According to a tribal legend, *"these hills were a reclining female figure from whose breasts flowed life-giving forces, and to them the Lakota went as a child to its mother's arms."*[152]

While the Black Hills have spiritual significance to several Indian tribes, they are most sacred to the Oglala Lakota people, who live on and nearby the Pine Ridge Reservation in southwest South Dakota. The reservation occupies all of Oglala Lakota County, the southern half of Jackson County, and Bennett County. The estimated total land area of the reservation is 2.1 million acres, with 1.7 million acres held in trust by the United States Government.[153]

Wounded Knee

The Pine Ridge Lakota are remembered because of their valiant chiefs such as Crazy Horse and Red Cloud. Pine Ridge was also the site of Indian militancy, including the Wounded Knee Massacre of 1890, where over 300 Lakota tribal members died, and the 1973 Occupation of Wounded Knee, which was a 71-day standoff between militants of the American Indian Movement (AIM) and federal law enforcement officials.

Wounded Knee is important to the Lakota and other American Indians because it is a *symbolic moment* in the relationship between Native Americans and White settlers. As a symbolic moment, Wounded Knee has spiritual significance. The Ghost Dance was a late 19th century

---

[151] Standing Bear, L., Land of the Spotted Eagle. Cambridge: The Riverside Press, 1933.

[152] Jewell, Benjamin, "Lakota Struggles for Cultural Survival: History, Health, and Reservation Life" (2006). Nebraska Anthropologist.

[153] Pine Ridge Indian Reservation, Re-Member website: https://www.re-member.org/pine-ridge-reservation.aspx, accessed September 8, 2019.

religious movement incorporated into many Native American belief systems, including the Lakota. Northern Paiute spiritual leader Wovoka had a vision that the proper practice of the dance could reunite the living with the spirits of the dead, bring the spirits to fight to defend Indian land rights and cultural preservation. He envisioned the Ghost Dance as a way to *"scare off"* White colonists, and bring peace, prosperity, and unity to Native American people.

Photograph 2: Wounded Knee Memorial Cemetery[154]

The practice of the Ghost Dance was believed to have contributed to the Lakota resistance to assimilation under the Dawes Act in the Wounded Knee Massacre. The Dawes Act of 1887 is described as: *"An Act to provide for the allotment of lands in severalty to Indians on the various reservations, and to extend the protection of the laws of the United States and the Territories over the Indians, and for other purposes."*[155] While the U.S. Government banned the Ghost Dance because it was seen as sparking Indian militancy, the practice went underground and continued

---

[154] Photograph by Don Iannone at Wounded Knee on the Pine Ridge Lakota Reservation on September 14, 2019.
[155] The Dawes Act, U.S. Congress: https://govtrackus.s3.amazonaws.com/legislink/pdf/stat/24/STATUTE-24-Pg387.pdf. Accessed January 9, 2020.

to be practiced by many Native American tribes. Many Natives Americans back then and today believe the underlying message of the Ghost Dance was one of pacificism, peace, and reestablishing harmony and balance to the world. Like prophecies in many religions, the Ghost Dance ceremony embodied a vision of a peaceful future life and world. With this understanding, Wounded Knee is sacred land for the Lakota as well as other Native Americans.

Brief Sketch of Lakota Quality of Life

What is the quality of life today of the Lakota living on Pine Ridge Reservation? Accurate American Indian reservation demographic, economic and health statistics are hard to find. Available data tend to paint only a partial picture of the sad realities endured by the residents of Indian communities like Pine Ridge.

The U.S. Census Bureau estimated that in 2017 that 19,779 Oglala Lakota Indians lived on the Pine Ridge Reservation.[156] The South Dakota Tribal Relations Office estimates that total enrolled membership in the Oglala Lakota tribe living everywhere was 38,332 in 2015. Life at Pine Ridge is characterized by a remarkably high poverty rate (53.75 percent), compared to the U.S. average of 15.6 percent. Educational attainment is very low with a school drop-out rate over 70 percent, and only 28.7 percent of the Native population of Pine Ridge Reservation has attained a high school diploma, GED or its alternative.[157] Oglala Lakota County ranked 59-out-of-60 counties in South Dakota for poor health outcomes in 2017.[158]

D. Lakota Spiritual Connection with the Black Hills

Cosmological Connection to the Black Hills

The Black Hills is considered to be sacred land by the Lakota people. Why are the Black Hills viewed as sacred land? According to Lakota mythology and cosmology, the Lakota people made a treaty a very long time ago with *Maka Wita* (the Star Nation or the Universe) to pray with a pipe and act as stewards of the sacred Black Hills, known as the

---

[156] My Tribal Area, U.S. Census Bureau at: https://www.census.gov/tribal/?aianihh=2810, accessed September 8, 2019.
[157] Pine Ridge Indian Reservation, Re-Member website: https://www.re-member.org/pine-ridge-reservation.aspx, accessed September 8, 2019.
[158] Ibid.

Pahá Sápa to the Lakota people. Within the Lakota spiritual worldview, the Black Hills consist of seven sacred mountains:[159]

1. Mato Tiplia (Bear's Lodge or Devils Tower in eastern Wyoming)
2. Mato Paha (Bear Butte in northwestern South Dakota)
3. Inyan Kaga (Stone Mountain in eastern Wyoming)
4. Pte Kinapapi (Wind Cave in western South Dakota)
5. Wakinyan Wahohpe (Thunderbird Nest in western South Dakota)
6. Pe Sla Paha (Bald Mountain in western South Dakota)
7. Pte Tatiopa (Buffalo Gap in western South Dakota)

These seven sacred mountains or sites in the Black Hills are seen as giving the Lakota people their consciousness. Within the Lakota worldview, consciousness refers to how the Lakota people see themselves (self-consciousness), the world (consciousness of others), and their relationships with the world (consciousness of how self and other are related). Through these seven sacred sites, the Lakota form a spiritual understanding of themselves and the universe. Therefore, the Black Hills are sacred lands to the Lakota.[160]

The Black Hills have always played a critically important role in Lakota cosmology. The anthropologist William K. Powers describes Lakota cosmology in this way: *"Cosmology, I think, like other aspects of that thing we call culture, should be seen as a viable, malleable entity, and like culture, cosmology provides the rationale for the manner in which humans adapt to their environments, including how they adapt to other cultures which are likewise part of the overall environment.*[161] Powers encourages us to view Lakota cosmology in dynamic terms, which means the Lakota see the universe as changing, which means as a people, the Lakota people are ever-changing.

Bison in the Lakota World

Bison have played an important role in Lakota culture and spirituality for many years. On a practical level, bison were the major source of food and materials for everyday life and spiritual ceremonies for

---

[159] White Lance, Frank, Why the Black Hills are Sacred: A Unified Theory of the Lakota Sundance, Ancestors Inc, Rapid City, SD, 2004, p. 29-36.
[160] Ibid, p. 59.
[161] Powers, William K. "Cosmology and the reinvention of culture: the Lakota case." The Canadian Journal of Native Studies 7.2 (1987): 165-180.

the Lakota for many years. In a spiritual sense, bison symbolize abundance and manifestation to the Lakota people. The major spiritual lesson here is that one does not have to struggle to survive if the right action is joined by the right prayer.[162] Bison also symbolize personal freedom to the Lakota, which is an extremely important tribal and personal value to the Lakota people.

The bison population, while almost decimated in the late 19th and early 20th centuries, is nearly 34,000 in South Dakota, according to a 2012 U.S. Census of Agriculture estimate. While the two largest publicly owned bison herds are found in Custer State Park (see Photograph 3 below) and the Badlands National Park, several thousand bison also live on tribal and private lands.

Photograph 3: Bison Herd in Black Hills in Custer, South Dakota[163]

---

[162] St. Joseph's Indian School, The Meaning of Tatanka and the Significance of the Buffalo to the Lakota People at:,https://www.stjo.org/buffalo-tatanka/. Accessed October 11, 2019.
[163] Photograph by Don Iannone in Custer State Park near Custer, South Dakota on September 11, 2019

THE BADLANDS AS SACRED LAND

A. Badlands Geological History

In the Badlands, the youngest geological formations are on the top and time marches straight downward, which is just the opposite of the geologic development of the Black Hills. The Badlands were formed primarily by two geologic processes (deposition and erosion), with the oldest exposed (and lowest) layers of the area being created 65-75 million years ago, when the surface area was covered by a warm inland sea.[164] When the hills uplifted, the sea drained away, and it was replaced by a river floodplain that deposited a new sediment layer every time a flood occurred. A drier period followed, bringing sediments deposited by the wind, with layer colors varying with time and volcanic activity. As the uplift to the west continued, the water current increased and carved into the very deposits earlier streams had left behind. Some 65 million years later, the eroded spires and valleys of the Badlands appeared.[165]

Where does the Badlands name come from? The Lakota people named this place *"mako sica,"* which means "land bad." Extreme high and low temperatures, lack of water, and the exposed rugged terrain led to this name. Similarly, French-Canadian fur trappers called the Badlands area the *"les mauvais terres pour traverse,"* which means *"bad lands to travel through."*[166] While the Badlands exhibit these characteristics, I experienced the 244,000-acre Badlands area as stunningly beautiful, especially in the early morning and late evening when the sun does its *"dance"* through the rocks, valleys, and surrounding grasslands. The sawtooth ridges and parched canyons inspired many generations of Native Americans, who embarked on vision quests for the Great Spirit to bestow good fortune on their people. The rugged, challenging nature of the Badlands made it a perfect place for vision quests and other ceremonies because it tested participants' courage, strength, and willingness to trust the spirits for deliverance and direction.

The Badlands National Park consists of three units. The North Unit is the location of the park headquarters and the Badlands Loop Road,

---

[164] Badlands National Park, Geologic Formations:
https://www.nps.gov/badl/learn/nature/geologicformations.htm, accessed September 21, 2019.
[165] Ibid.
[166] How Did the Badlands National Park Get Its Name?: https://www.blackhillsbadlands.com/blog/2014-07-10/how-did-badlands-national-park-get-its-name, accessed September 21, 2019.

which is nearly surrounded by Buffalo Gap National Grassland. The second two units are located in the south: The Stronghold Unit and Palmer Creek Unit, which are within the Pine Ridge Indian Reservation of the Oglala Lakota.

Photograph 4: Badlands Seen from Northern Pinnacles Entrance[167]

B. Badlands History, Ghost Dance, Spiritual Roots

Short Cultural History

Human occupation and hunting in the Badlands date back over 11,000 years. Paleo-Indians, the Arikara (now in North Dakota), and more recently the Lakotas, have a colorful history in the Badlands. Clashes between homesteaders and American Indians were common in the area during the past 200 years.

Ghost Dances in the Badlands

The Ghost Dance was discussed earlier, but its mention in relation to the Badlands is important. The Lakota Ghost Dance Ceremony originated in the Badlands in 1890, and it culminated in the Battle of

---

[167] Photograph by Don Iannone in Badlands National Park, Pinnacles Entrance, near Wall, South Dakota on September 13, 2019

Wounded Knee Creek, which is about 45 miles south of the Badlands. The circular dance was believed to promote the return of the dead and the restoration of traditional ways of Indian life. With the forced Christianization of American Indians during the late 19th and early 20th centuries, as mentioned earlier, the Ghost Dance and other traditional Indian ceremonies and dances were prohibited as *"pagan"* religious activities by the U.S. Commissioner of Indian Affairs.[168]

For many years, the Lakota collected fossilized bones, shells, rocks, and plants in the Badlands for use in their spiritual ceremonies. The shells are seen as reminders that once a sea existed where the rocky spires of the Badlands now stand.[169] The Stronghold area, or the *"Oonakizin,"* of the Badlands is a sacred place of strength for the Lakota. This is the southern unit of the Badlands National Park, which is located on the Pine Ridge Reservation. Many Lakota prayer bundles are found tied to tree and shrub branches in this area. Some Lakota believe the last Lakota Ghost Dance of the late 19th century was in the Stronghold area.

The fossil remains of life from millions of years ago are considered sacred to the Lakota.[170] These fossils speak not only to earlier times but to the timeless quality of the Badlands. In this sense, the Lakota see not only rock and sky, but they see their ancestors and the great trail of beings that gave birth to today's living. The Lakota consider the paleontological and archeological digs in the Badlands as desecrations of sacred land. Lakota vision quests in the Badlands, especially in the Stronghold area, remain quite common today. Finally, the Badlands, like other Lakota sacred lands, are seen as filled with and animated by the Great Spirit.

Bighorn Sheep in the Badlands

The bighorn sheep seen at various locations in the Badlands have symbolic meaning to the Lakota. They represent the ruggedness of life, or the necessity of surviving life's struggles. The sheep also represent sure-footedness, or the need to walk through life with balance and confidence. Finally, these animals symbolize sacrifice in the Indian world, where they are killed in sacred ceremonies. Photograph 5 below shows the bighorn

---

[168] DeMallie, Raymond J. "The Lakota Ghost Dance: An Ethnohistorical Account." Pacific Historical Review 51.4 (1982): 385-405.
[169] Personal communication with Roger Broer, a prominent artist, Lakota elder and enrolled member of the Oglala Lakota Nation on September 3, 2019.
[170] Benton, Rachel, C., et al. "Baseline mapping of fossil bone beds at Badlands National Park." 6th Annual Fossil Resources Conference. 2001.

sheep on a Badlands ridge. During our field visit to the Badlands on September 13, 2019, we saw the sheep in several parts of the park.

It is believed that 200 years ago bighorn sheep were widespread throughout the western United States, Canada, and as far south as northern Mexico. Some estimates placed their population at over 2 million. By around 1900, hunting, competition from ranching, and diseases had decreased the population to several thousand.[171] The bighorn sheep's main threats are unregulated or illegal hunting, diseases introduced from outside their residential areas, competition from livestock, and habitat encroachment by humans.

Photograph 5: Bighorn Sheep in the Badlands[172]

---

[171] Bighorn Sheep (Ovis canadensis): https://www3.northern.edu/natsource/MAMMALS/Bighor1.htm. Accessed January 9, 2020.
[172] Photograph by Don Iannone in Badlands National Park, Pinnacles Entrance, near Wall, South Dakota on September 13, 2019

In Sacred Relationship

DEVILS TOWER AS SACRED LAND

A. Introduction

Devils Tower is sacred ground to the Lakota and other western American Indian tribes. Before the arrival of White colonists, the Lakota, Shoshone, Kiowa, Crow, and Cheyenne native tribes worshipped at Devils Tower. Archeologists have found evidence that Native Americans have been living in the vicinity for at least 10,000 years.

Devils Tower is located near Hulett, Wyoming. It is called *"mato tipi la paha"* or *"The Hill of the Bear's Lodge"* by the Lakota. More often, Devils Tower is called *Mato Tipila*, or Bear Lodge. The Lakota have shared their knowledge of the stars across generations. The Lakota *"Mato Tipila"* star constellation shines brightly in the northern hemisphere. The fact that this constellation shares the same name as the geologic feature known as Devils Tower is no coincidence.

Devils Tower was brought to public attention by the hugely popular 1977 science fiction movie, *Close Encounters of the Third Kind,*[173] which was filmed in part at Devils Tower. Third Kind encounters refer to human contact with aliens.

B. Devils Tower Geological History

The oldest rocks visible in Devils Tower were deposited in a shallow inland sea. This sea covered much of the central and western United States during the Triassic period, which was 250 to 200 million years ago. Dark red sandstone and maroon siltstone, interbedded with shale, can be seen along the Belle Fourche River. Oxidation of iron-rich minerals causes the red color of the rocks.

Many geological studies conclude that Devils Tower began as magma, or molten rock buried beneath the Earth's surface. A remaining mystery is what processes caused the magma to cool and form the Tower. One explanation is that Devils Tower is a stock, which is a small intrusive body formed by magma which cooled underground and was later exposed by erosion.[174]

---

[173] Close Encounters of the Third Kind Movie website: https://www.imdb.com/title/tt0075860/. Accessed September 6, 2019

[174] National Park Service, Devils Tower: https://www.nps.gov/deto/index.htm. Accessed September 22, 2019.

In Sacred Relationship

Another theory, which lacks demonstrable evidence such as volcanic ash, lava flows, and volcanic debris, suggests that Devils Tower is a volcanic plug or the neck of an extinct volcano. The columns of Devils Tower are its most striking feature. This geologic feature is known as columnar jointing. Geologists believe that the Tower formed around 50 million years ago. At that time, it was one to two miles below the Earth's surface. Between 5 and 10 million years ago, erosive forces began to expose the Tower, making it visible as we see it today.

Devils Tower is 1,267 feet above the nearby Belle Fourche River, but from the base to the top it is 867 feet. Devils Tower's elevation is 5,112 feet above sea level. Photographs 6 and 7 below show Devils Tower from a distance and close-up. Devils Tower was the very first official United States National Monument, which was proclaimed in 1906 by President Theodore Roosevelt, who loved the American West, shortly after he signed the American Antiquities Act into law.[175] [176]

Photograph 6: Devils Tower Seen from a Distance[177]

---

[175] Ibid.
[176] National Park Service, Devils Tower Geology: https://www.nps.gov/deto/learn/nature/tower-formation.htm. Accessed September 22, 2019.
[177] Photograph of Devils Tower seen from a distance by Don Iannone on September 13, 2019.

Photograph 7: Devils Tower Seen Close-Up[178]

C. Cultural and Spiritual Connection to Devils Tower

The earlier section provided a scientific account of how Devils Tower formed. The Lakota explanation is quite different, according to Jeffrey Ostler, a history professor at the University of Oregon. Ostler shared this Lakota story about the genesis of Devils Tower: "*Their name for Devils Tower is Mato Tiplia (Bear's Lodge) and they say that Bear's Lodge was created when several girls who wandered away from their camp were chased by bears. Suddenly, the earth rose, carrying the girls out of reach as the pursuing bears clawed frantically at the rising earth, creating he long crevices. The girls were rescued by the birds, sent by Fallen Star, one of the Lakotas' cultural heroes.*"[179]

Devils Tower remains an important sacred place for the Lakota people. Ostler says the Lakota use two places for their Sun Dance, which was their great *corporate* prayer. These places are Sundance Mountain and Devils Tower. The Sun Dance was the most important ceremony

---
[178] Photograph of Devils Tower seen close-up by Don Iannone on September 13, 2019.
[179] Ostler, Jeffrey, The Lakotas and the Black Hills, Struggle for Sacred Ground, Penguin Books, NY, 2010, p.5.

practiced by the Lakota and nearly all Plains Indians. The Sun Dance was a time of renewal for the Lakota people and the earth itself. During the Sun Dance ceremony, the Lakota reaffirm their basic beliefs about the universe and the supernatural through rituals of sacrifice.[180] [181]

The Lakota still fast, pray, leave offerings, worship the *"Great Mystery,"* and perform sweat lodge ceremonies at Devils Tower. Lakota healing ceremonies are held there, but because of the many non-Indians visiting Devils Tower, these ceremonies are less frequent there. Perhaps the most controversial issue to the Lakota and other American Indian people is the use of site for recreational climbing by tourists, which my wife and I observed during our Devils Tower visit. Devils Tower is considered as a birthplace of Lakota wisdom. Lakota vision quests were common in earlier days at Devils Tower. It is believed the Lakota received the White Buffalo Calf Pipe, the most sacred object of the Lakota, at Bear Lodge. While Buffalo Calf Woman is a legendary spiritual being to the Lakota. The sacred pipe's sanctuary was found within a secret cave on the north side of Bear Lodge.[182]

OTHER LAKOTA SACRED LANDS AND SITES

A. Bear Butte

Bear Butte is located near Sturgis, South Dakota, which is in the northwest corner of the Black Hills. The mountain resembles a huge sleeping bear, which explains its name. Bear Butte is not strictly a butte. It was created primarily by erosion of sedimentary strata. Bear Butte is instead a laccolith, or an intrusive body of igneous rock, uplifting the earlier sedimentary layers, which have since largely eroded away.[183] The Cheyenne named the site *"Noahavose"* meaning the Good Mountain, and the Lakota named it *"Pahan Wakan"* meaning Bear Mountain. This stand-alone butte was a powerful sacred site used for vision quests that was used by the Plains Indians for many millennia. It was on Bear Butte that Crazy Horse had his famous vision for the future of all Indians would

---

[180] Ibid, p. 17-19.

[181] Powers, William, K., Oglala Religion, First Bison Book Printing, November 1982, p. 95-100.

[182] Hanson, J. R., and S. Chirinos. 1991. "Ethnographic Overview and Assessment of Devils Tower National Monument." University of Texas, Arlington.

[183] Mato Paha Bear Butte: http://aktalakota.stjo.org/site/News2?page=NewsArticle&id=8999. Accessed September 27, 2019.

experience hard times ahead, then eventually arrive at a period of spiritual awakening and peace.

B. Wind Cave

According to Lakota legend, there is a sacred cave called Wind Cave in the Black Hills, which was the source of bison herds that once roamed the Plains in great numbers. A steady, constant wind from the cave's inner reaches is said to have blown the herds out from under the earth to feed the Lakota people.

Wind Cave is in the southwestern corner of South Dakota. It is known for its vast underground chambers. The cave's walls are rich in honeycomb-shaped calcite formations called boxwork.[184] The National Park Service says: *"Wind Cave is old, complex, and filled with more boxwork than is found in all other caves on Earth put together. Any one of these qualities would make Wind Cave unique. Together they make it a world-class cave. And each part is essential to understanding how the cave formed."*[185]

---

[184] Wind Cave National Park: https://www.nps.gov/wica/index.htm. Accessed September 30, 2019.
[185] National Park Service, Wind Cave: https://www.nps.gov/wica/learn/nature/wind-cave-geology.htm. Accessed December 4, 2019.

CHAPTER 6: LAKOTA SPIRITUAL WORLDVIEW AND SACRED LAND

"Wakan Tanka, Great Mystery, teach me how to trust my heart, my mind, my intuition, my inner knowing, the senses of my body, the blessings of my spirit. Teach me to trust these things so that I may enter my sacred space and love beyond my fear, and thus walk in balance with the passing of each glorious Sun."
Lakota Prayer

A. Introduction

This chapter examines the relationship between the Lakota spiritual worldview and sacred land. This relationship is examined from three standpoints: 1) how geographic place has influenced the Lakota spiritual worldview; 2) how the Lakota spiritual worldview has influenced how the Lakota view geographic areas and sacred places; and 3) how sacred meaning is attached to specific places by the Lakota. Chapter 6 draws heavily upon this book's earlier chapters, the published literature, and personal interviews I conducted in September 2019 with Lakota scholars, teachers, and elders.

B. Place Influences the Lakota Spiritual Worldview

Geography influences how we see the world, including our spiritual worldview.[186] From the standpoint of visual perception, geography and place shape what we think and feel. The mountains, valleys, skies, unique rock formations, forests, rivers, creeks, rivers and waterfalls, fields, pastures, and plains, and wildlife of western South Dakota and eastern Wyoming evoke feelings of beauty, peace and calm, inspiration, and awe and wonder. In many cases, they give rise to feelings of a holy or divine

---

[186] Kong, L. (1990) Geography and religion; trends and prospects. Progress in Human Geography 14; 355-371

presence. In this respect, it is understandable why the Lakota people for many years have been attached to these places in a deeply spiritual sense.

In understanding how geography and place influence spiritual beliefs, culture, and human behavior, it is important to view geography and place as influences rather than unchangeable determinants. The latter is called environmental determinism, also known as geographical determinism, which is a theory about how the physical environment predisposes people, cultures, and societies toward particular location and development patterns.[187] Some scholars have concluded that environmental determinism ignores the influence of human action and beliefs on cultures and societies. Moreover, they say that environmental determinism has been used as a justification for colonialism, imperialism, and eurocentrism (favoring the Western worldview).[188] [189] [190]

The geographer C. G. Glacken sees the relationship between geography and religion as having deep historical roots. Glacken writes: *"In ancient and modern times alike, theology and geography have often been closely related studies because they meet at crucial points of human curiosity. If we seek after the nature of God, we must consider the nature of man and the earth, and if we look at the earth, questions of divine purpose in its creation and of the role of mankind on it inevitably arise."*[191] Glacken's perspective is consistent with how Lakota spirituality has been influenced by geography.

The religious historian Philip Sheldrake reminds us that place is something different, and more, than geography: *"We come to know reality in terms of our experience of specific places. What does 'place' mean? It involves far more than geography. 'place' is location with particular significance because of its connection with the people who live 'here' rather than somewhere else, or because it evokes something of significance, for example the historical memories of a long-standing community. 'Place' thus involves a dialectical relationship between*

---

[187] Conceptually (January 20, 2019). "Determinism - Explanation and examples". conceptually.org, accessed September 27, 2019.
[188] Gilmartin, M. (2009). "Colonialism/Imperialism". Key concepts in political geography, (pp. 115–123). London: Sage Publishing
[189] Sluyter, Ander (2003). "Neo-Environmental Determinism, Intellectual Damage Control, and Nature/Society Science". Antipode-Blackwell.
[190] Rickett, Allyn, in Guanzi: Political, Economic, and Philosophical Essays from Early China: A Study and Translation. Volume II. Princeton University Press, 1998, p. 106.
[191] Glacken, C., J., Traces on the Rhodian Shore. Berkeley, CA: University of California Press, 1967, p. 35

physical environments, including cities, and human narratives."[192] Sheldrake continues by saying that place is characterized by *"particularity,"* which signifies the personal experiential dimension of place.

The Lakota see the Black Hills, the Badlands, and Devils Tower as very special places that are spiritually significant in part because of Lakota creation stories connecting the Lakota to these specific places, and because of the Lakota people's longstanding personal experiences with these places. Some of these creation stories were discussed earlier.

Many American Indians see the land and sacred sites as having been given to them by their Creator. Winona LaDuke, an environmental activist and writer with paternal roots in the Ojibwe White Earth Reservation in Minnesota, explains this phenomenon in the following way: *"Since the beginning of time, the Creator and Mother Earth have given our peoples places to learn the teachings that will allow us to continue to be, and to reaffirm our responsibilities and ways to be on the lands from which we have come. Indigenous peoples are place-based societies, and at the center of those places are the most sacred of our sites, where we reaffirm our relationships. Everywhere there are indigenous people, there are sacred sites, there are ways of knowing, there are relationships."*[193]

From the traditional Lakota spiritual perspective, everything in the world, including the land, is seen as alive or living. The Lakota believe they share a spiritual connection or relationship with the land and the places they know. For this reason, the Lakota honor these lands and places because of this relationship. This relationship is based in *animism*, which has received considerable research attention, especially in anthropology, shamanism, and earth-based spirituality studies. Graham Harvey says there are two kinds of animism. The first is rooted in a concern with what is alive and what makes a being alive. This is the older view, which has been challenged on a scientific basis as believing in spirits or non-empirical beings. The second view, which is newer, refers to a concern with knowing how to behave appropriately towards *"persons,"* not all of which are human. The real concern in the second understanding of animism is how beings are to be treated or acted towards.[194]

Anthropologist David Posthumus sees animism playing a

---

[192] Sheldrake, Philip, The Spiritual City: Theology, Spirituality, and the Urban, First Edition. John Wiley & Sons, Ltd., 2014, p. 117-118.

[193] LaDuke, Winona, In the Time of the Sacred Places in The Wiley Blackwell Companion to Religion and Ecology, First Edition. Edited by John Hart, 2017, John Wiley & Sons Ltd. p. 73-74.

[194] Harvey, Graham. Animism: Respecting the living world. Wakefield Press, 2005, p. xi.

significant role in Lakota spirituality. He quotes Lakota spiritual leader Luther Standing Bear:[195] *"From Wakan Tanka there came a great unifying life force that flowed in and through all things—the flowers of the plains, blowing winds, rocks, trees, birds, animals—and was the same force that had been breathed into the first man. Thus, all things were kindred and brought together by the same Great Mystery. Kinship with all creatures of the earth, sky, and water was a real and active principle."*[196] Posthumus continues to say, *"This animist worldview effectively dissolved the boundary between what Westerners call nature and culture."*[197] Interviews with Lakota elders and spiritual teachers affirm this view that there is no separation between *"nature"* and *"culture."*[198] Similarly, there is no boundary between the realities of physical and spiritual existence in the past or present. It is believed that the spiritual beings animating the Black Hills, the Badlands, and Devils Tower in the past still animate these places today.[199]

Animism is akin to the concepts of genius loci, spirit of place, and terrapychology. Brian Spittles writes: *"On the whole, genius loci is used in the fields of architecture, landscape design and the arts to signify a human sense of the essence of external environments. Spirit of place, however, is more often used in literature in reference to the pervasive numinous 'spirit' of natural places, or in the words of D. H. Lawrence, 'different places on the face of the earth have different vital effluence, different vibration, different chemical exhalation, different polarity with different stars: call it what you like. But the spirit of place is a great reality.'"*[200]

What is Terrapsychology? *"Terrapsychology explores how terrain, place, element, and natural process show up in human psychology, endeavor, and story, including myth and folklore. Mainstream psychology treats people like machinery while ignoring our connection to nature, place, and planet. By contrast, terrapychology aligns itself with ecology and the humanities to re-story our place in a world filled with wonders and*

---

[195] Posthumus, David C. "All My Relatives: Exploring Nineteenth-Century Lakota Ontology and Belief." Ethnohistory 64.3 (2017): 379-400.

[196] Standing Bear, Luther, 2006, Land of the Spotted Eagle. Lincoln: University of Nebraska Press.

[197] Posthumus, David C. "All My Relatives: Exploring Nineteenth-Century Lakota Ontology and Belief." Ethnohistory 64.3 (2017), p. 383.

[198] For many American Indians, the worlds of nature and culture are connected by spiritual beings and forces, which gives them divine oneness or wholeness.

[199] Personal communications with Jace DeCory (Cheyenne River Lakota) at Black Hills State University on September 12, 2019 and Roger Broer (Oglala Lakota) on September 3, 2019.

[200] Spittles, Brian. "Making sense of a sense of place." Colloquy 30 (2015), p. 128.

living presences we are always in conversation with."[201]

C. Lakota Worldview Influences the View of Place

As humans, what we feel, think, and believe influences how we view and treat geographic places. From this perspective, our spiritual worldviews have a bearing on how we see larger geographic areas and specific places within these areas. Spiritual worldview influences the Lakota view of and interaction with geographic places in six ways:[202] [203] [204]

1. **Principle of Oneness**: Traditionally, the Lakota see themselves as one (in unity) with their land and sacred places. In this sense, there is no separation between people and land and sacred places. Through their belief in animism, or the belief in a supernatural power that organizes and animates everything in the physical world, the Lakota see and experience a deep spiritual connection with their sacred land and sacred places.

2. **All Things are Contained in Nature**: The Lakota see themselves and everything else in life as belonging to nature. In this respect, humans exist within and not outside nature.

3. **Moral Stewardship of the Earth**: Lakota spiritual teachings encourage moral stewardship of the Earth and Lakota sacred places. This moral stewardship is a natural outflow of the beliefs in oneness and all things are contained within nature.[205]

4. **Ancestral Spiritual Ties**: Because the Lakota see the living and dead as eternally connected in the spirit realm, each generation of Lakota people is connected to sacred places through their ancestral ties. Kinship ties to humans and non-humans are primary to the Lakota way of life. These ties lead to the Lakota view of *"all are relations."* This view is common to many other

---

[201] What is Terrapychology? At: http://www.terrapsych.com/, accessed October 1, 2019.
[202] Personal communication with David Posthumus on September 4, 2019 and personal communication with Jace DeCory on September 12, 2019.
[203] Powers, William, K., Oglala Religion, Bison Book Printing, 1982, p. 44-55.
[204] Deloria, Vine, Spirit and Reason, Fulcrum Publishing: Golden, CO, 1999, p. 56-58.
[205] Ibid, p. 323-338.

Native Americans, including the Western Apache people in Arizona.[206] The Lakota care for the land because the land cares for them.

5. **Earth as Medicine**: The Earth, including sacred lands, provide the *"medicine"* the Lakota need to live in wellness, balance, and harmony. This includes native food, plant medicine, and even the rocks and earth itself. In this sense, the bison is a sacred creature sustaining the Lakota in terms of nourishment of mind, body, and spirit. The medicine wheel, discussed earlier, is a symbol of *"Earth as medicine."*[207]

6. **As Above, So Below**: Cosmology influences how the Lakota see geographic areas and sacred places in these areas. *Lakota Star Knowledge* provides a cosmological explanation of how Lakota sacred lands and sites emerged through their alignment with the stars and planets in the heavens. Native American spiritual teacher Paula Giese explains, *"What is in the stars is on earth and what is on earth is in the stars. This idea unites the ceremonial map in the circle of stars, not only with sites in the Black Hills, but with a round of ceremonial actions at sacred sites there, ending with a Sun Dance at the Bear's Lodge (Devil's Tower)."*[208]

D. How Sacred Meaning is Attached to Place

This issue was addressed earlier. It is explored in more depth here. According to Chidester and Linenthal, there are three ways sacred meaning is attached to specific places. The first way is through the identification with special spiritual powers or energies inherent in certain places or lands, which cause them to be seen as sacred in nature. This was described in the earlier section as animism, which is a part of the Lakota spiritual worldview.

As Zimmerman and Molyneaux say in their book *Native North America*: *"In Native belief, animals have spirits, just like human beings,*

---

[206] Basso, Keith, H., Wisdom Sits in Places, University of New Mexico Press, 1996, p. 3-8.

[207] Zimmerman, Larry, J., and Molyneaux, Brian, Leigh, Native North America, University of Oklahoma Press: Norman, 2000, p. 76-77.

[208] Giese, Paula Lakota Star Knowledge, 1995 at: http://www.kstrom.net/isk/stars/startabs.html, accessed September 29, 2019.

In Sacred Relationship

and enjoy a complex reciprocal relationship with people, plants, and the earth."[209] The authors go on to say that each Native culture has special reverence for the region in which they live. Carson McCullers wrote in *The Heart Is a Lonely Hunter*, "*To know who you are, you have to have a place to come from.*"[210] This is very much the idea underlying how and why sacred meaning gets attached to the specific places we call our *"homes"* in the sense of everyday life and spiritual existence.

The second approach to attaching sacred meaning to geographic place is by *"sacralization,"* which uses cultural ritual. This reminds us of the importance of the reenactment of creation stories through spiritual ceremonies. Basso quotes Leslie Marmon Silko about the critical connection between narratives and the land and places: "*The stories cannot be separated from geographical locations, from actual physical places within the land...And the stories are so much a part of these places that it is almost impossible for future generations to lose the stories because there are so many imposing geological elements...you cannot live in that land without asking or looking at or noticing a boulder or rock. And there's always a story.*"[211]

Sacred places in Lakota spirituality are spaces where the *"veil"* between humans and the transcendent and eternal is thin, which allows the communing of the spirit world and the earth plane world. Rituals, including the Sun Dance, sweat lodge, medicine wheel ceremonies, and smoking the sacred pipe, are believed to bridge the spiritual and earthly worlds. The land itself is seen as a connective force uniting the living, dead, and eternal. This connective force is enlivened and experienced by the Lakota through ritual and sacralization.[212]

The third way to attach sacred meaning to place combines the *"inherent spiritual power present within the place or space"* approach and the *"sacralization through ritual and ceremony"* approach.[213] As mentioned earlier, Great Spirit's eternal presence and animism are seen as imbuing certain places with sacredness. In this sense, sacredness already exists, which is accessed by the Lakota through ritual and ceremony. The origin,

---

[209] Ibid, p. 78.

[210] McCullers, Carson. The heart is a lonely hunter. Houghton Mifflin Harcourt, 2010.

[211] Basso, Keith, H., Wisdom Sits in Places, University of New Mexico Press, 1996, p. 64, and Silko, Leslie Marmon. Language and literature from a Pueblo Indian perspective. na, 1981, p.69.

[212] Deloria, Vine, Spirit and Reason, Golden, CO: Fulcrum Publishing, 1999, p. 323-337.

[213] Chidester, David, and Edward Tabor Linenthal, eds. American sacred space. Indiana University Press, 1995, p. 5-7.

In Sacred Relationship

or beginning, of this sacredness is often explained by a special event that occurred in a particular place many years ago. All Lakota creation stories took place somewhere specifically. These specific places are for this reason *"sacred land."*[214]

Finally, the Lakota believe that everyone has a moral responsibility to honor the Great Spirit and treat the land with respect and gratitude. This ecological ethic is a second explanation for how inherent spiritual presence and ceremony and ritual are brought together. The notion of *"walking lightly on the Earth and leaving few footprints"* is important to the Lakota and many other American Indians.

---

[214] Deloria, Vine, Spirit and Reason, Golden, CO: Fulcrum Publishing, 1999, p. 327.

PART III:
WORKING WITH GROUNDED SPIRITUALITY AND SACRED RELATIONSHIPS

*"Only when you are fully engaged can you see the activity
that will make your life feel worth living."*
~ Thomas Moore, Psychoanalyst and Author

CHAPTER 7: THE GROUNDED SPIRITUALITY AND SACRED RELATIONSHIP COMPASS©

"In talking to children, the old Lakota would place a hand on the ground and explain: 'We sit in the lap of our Mother. From her we, and all other living things, come. We shall soon pass, but the place where we now rest will last forever.' So, we, too, learned to sit or lie on the ground and become conscious of life about us in its multitude of forms."
Lakota Chief Luther Standing Bear

A. Caveat

The spiritual compass model presented in this chapter is an outgrowth of my experience with Lakota spirituality, but it goes beyond to include ideas from interfaith spirituality, interspirituality, and other wisdom sources. *Interfaith spirituality* is inclusive by nature, by honoring all religions and spiritual traditions. Interfaith is a *way of life* that allows each person to experience the Divine in their own way. *Interspirituality* is about the *deeper unity of spiritual experience*, which underlies all religions and spiritual traditions. The overall intent *of In Sacred Relationship* is to provide a grounded spirituality and sacred relationship compass (GSSRC), which any reader can use as a tool within their own religion or spiritual tradition.

B. Grounded Spirituality and Sacred Relationships Definitions

The Soul and Its Imaginative Possibilities

Lakota spirituality teaches us about the importance of our relationships with the whole of life, and it presents opportunities for us to grow our sacred relationships with ourselves, others, the Earth, our community, and the Great Spirit or God. As human beings, we share many similarities, and yet each of our lives is created in a unique *"image"* of our soul and God. This image grounds us, and it helps us find our way through this lifetime, and quite possibly other lifetimes of the soul. I believe in the reality of the soul as our immortal spiritual essence, which manifests in several lifetimes. As the reader of this book, you are free to *"imagine"* the soul in any way that makes sense to you.

> Lakota spirituality teaches us to look carefully at "all our relations," which includes our relationship with ourselves, other people, our community, Mother Earth, the universe, and Great Spirit. How we relate in the world gives definition to who we are and how we are in life. Grounded spirituality helps us form the right relations in all these areas.

How might we understand and encounter the soul? The archetypal psychologist James Hillman says: *"the soul represents the imaginative possibility of our nature, a possibility that is realized in reflective speculation, dream, image, and fantasy."*[215] Hillman continues to say that: *"by entering the imagination we cross into numinous precincts. And from within this territory, all events in the soul require religious reflection."*[216] The phrase *"imaginative possibility of our nature"* encourages us to see the soul in dynamic terms, and to view the soul as the source of our imagined possibilities in life. Hillman is trying to help us see the *"creative"* role played by the soul in our lives. Rather than dwelling on what is the nature of the soul, Hillman says we should focus on how and what the soul manifests in our life. When we are *in sacred relationship* with our soul, these manifestations are meaningful to the soul, which align the purposes of our soul and God.

Archetypal psychology (soul psychology) teaches us about the power of images, especially images arising from the soul, which includes each of us as an *"image"* created by the soul. For this reason, what we imagine about our soul, and the images our soul sends us, shape our relationship with our soul and everything else. Human beings relate through images, even more so than words. The relationships we form with these images either grounds us in a positive way or disconnects us from our soul. Hopefully, they ground us. *Lakota spirituality reminds us of the sacred territory inside each of us, and this inner sacred territory's need to connect with the sacred layers of reality surrounding it, including our*

---
[215] Hillman, J., Re-Visioning psychology. New York: Harper & Row, 1975, p. 112.
[216] Ibid, p.226.

bodies, the earth, community, and the Great Spirit or God. This strikes me as just what a religion or spiritual tradition should do! This is what many people truly hunger for in their lives, which is a deeper sustained spiritual connection.

Grounded Spirituality

What is grounded spirituality? Grounded spirituality is spiritual faith and practice that help us understand and manifest our lives in line with our souls. Most importantly, grounded spirituality is a spiritually centered *way of life*, or a way of living that encourages us to seek balance, harmony, and wholeness in our lives. Grounded spirituality recognizes our intertwined human and divine natures, and the ongoing challenge of aligning the *"relationship"* between the two. Grounded spirituality is contemplative in nature, encouraging us to form an awareness of our integral connection (relationship) with the vast and dynamic web of life. This awareness expands our sense of ourselves and the world. In short, grounded spirituality consists of spiritual beliefs and practices that help us experience the *gift of life* and its supreme beauty. *Finally, grounded spirituality is practical spirituality that encourages us to be who we really are, and be the best at that each day.* It urges us to practice our beliefs. In that way, we help our beliefs to work for us.

Connecting Grounded Spirituality and Sacred Relationships

Let us look at how grounded spirituality and sacred relationships fit together. Grounded spirituality sets the stage for our religion or spirituality to serve us in everyday life. It gets us away from thinking about religion and spirituality as something separate from other aspects of life. Grounded spiritualty is a *way of life*, as mentioned earlier. It empowers and enables us to approach life in a spiritually grounded way. Sacred relationship-building is an outgrowth of grounded spirituality, and it is the primary way we relate to life each day. When we live in sacred relationship with our souls, other people, our bodies, and the earth, we are on a grounded path to bring harmony, balance, and wholeness to our lives.

Here is an example of how the two fit together. In the Christian and Jewish faiths, the Ten Commandments are essential guides to living good Christian and Jewish lives. For the Ten Commandments to act as *real* guides in our lives, we must apply them in meaningful ways to how we live. Because each of us has a unique soul, personality, and life experiences, the Ten Commandments are not abstractions, they mean something _specific_ to each of us. The recognition of this specific significance is what makes the Ten Commandments personally important and meaningful. The Ten Commandments become _grounded_ spiritual beliefs when we live the Commandments in a real way; that is when they become a part of how we live our lives as Christians or Jews. Additionally, each of the Ten Commandments requires us to form sacred relationships. In the case of the Fifth Commandment, which is _"Honor thy father and mother,"_ it becomes a sacred relationship when we honor our parents as we honor God. Our birth, which is among the most sacred events in life, is possible because of our parents and God working through them. The Fifth Commandment asks us to honor the relationship with our parents as sacred because of God's role in our creation. _As I said earlier, a primary goal of this book is to provide a spiritual compass that can be adapted to any religion or spiritual tradition._

C. Grounded Spirituality and Sacred Relationships: A New Spiritual Compass

This chapter presents a model for cultivating grounded spirituality and sacred relationships in today's world. As shown in Figure 6 below, the model has four primary realms. _As suggested earlier, please view the model as a "compass" that helps you navigate the four primary realms of your life._ The four realms are:

1. The Divine Realm (God, Creator, Higher Power, and Soul);
2. Your Personal Realm (Spirit, Mind, and Body);
3. Realm of Others (People individually and collectively); and
4. Whole of Life Realm (All sentient beings and all things).

The Grounded Spirituality and Sacred Relationship Compass (GSSRC) is intended as a _living model that we can grow into._ The compass' realms are interrelated, which means we find ourselves in all realms at the same time. The GSSRC, as a spiritual compass, is designed to help us navigate these realms each day of our lives. Some days we are in close sacred relationship with our personal realm, while other days our spiritual energy is tied to the realm of others. The value of the compass is its invitation for us to be aware of how we navigate these realms of life.

In Sacred Relationship

Quite importantly, the intention of the GSSRC is to bring out our personal spiritual beauty, and not set us in motion to find some abstract external ideal. Our journeys with the compass are inner journeys, which can connect us in the right ways with the outer world.

My advice to readers is to contemplate the compass a bit, look at your life, and find how the compass can work for you <u>now</u>. Once again, each of us is different. Find an accessible starting point and work from that. Do not spend too much time searching for a *"perfect"* starting place. There isn't one. In my own experience, if we work at grounding ourselves in our bodies, with the earth, and with others, we encounter the Divine at work. The soul, as an aspect of the Divine, must be approached indirectly, as Parker Palmer reminds us.[217] Palmer says therefore art, music, natural beauty, mystery, silence, walks in the woods, journaling, heartful conversations, and many other activities promote soulfulness. In general, paying attention to the intricate details of life is an effective way to connect with our souls. Our connection with these details allows us to experience the richness of life. Soulfulness and richness go hand in hand.

Figure 6: The Grounded Spirituality and Sacred Relationship Compass (GSSRC)©

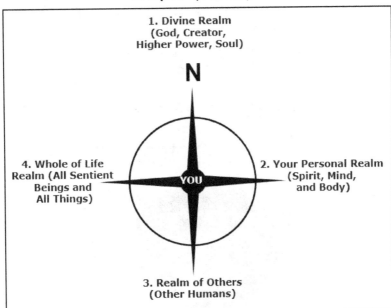

---

[217] Ibid

The grounded spirituality and sacred relationship compass (GSSRC) places *"You"* in the center of the compass, and your sacred relationships with the Divine, Your Personal Realm (Spirit, Mind, and Body), Other People, and the Whole of Life. When we are in sacred relationship with the Divine, we are in alignment with our *"Spiritual True North."* The GGSRC is a *"medicine wheel"* of sorts, but its directions and characteristics are not the same as those associated with the Lakota Medicine Wheel.

First Realm: Sacred Relationships with the Divine

A. Divine Realm Overview

Many names are used for the Divine in world religions and spiritual traditions, such as God, The Creator, and The Higher Power. In addition to religious and spiritual tradition definitions of the Divine, everyone has personal ideas about the Divine. To ensure that the GSSRC has value to people of different faiths, the name *"Divine"* is used in the compass.

Briefly, I would like to share some of my personal spiritual beliefs, with no intention of changing your beliefs as a reader of this book. I have arrived at my beliefs after many years of spiritual self-work. Earlier I introduced the concepts of interfaith spirituality (spiritual diversity), which connects across faiths and interspirituality (spiritual unity), which runs through faiths. Please see Figure 7 below.

Figure 7: Visualization of Interfaith Spirituality and Interspirituality

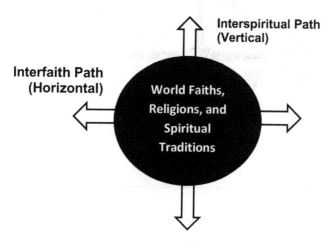

The Divine is the spiritual realm, which is the *spiritual home* to which our spirits return upon our physical death. The soul belongs to the Divine Realm, and therefore it is our immortal spiritual essence. The soul dwells in the spiritual realm, and not within a person. The spirit, on the other hand, resides within (animates) the person. At physical death, the spirit returns to the spiritual realm and rejoins its soul. The soul, as an immortal spiritual essence, has many spiritual lives in human form. The soul belongs to God, and the spirit belongs to the soul.

Figure 8: Soul, Spiritual Realm, and Human Realm

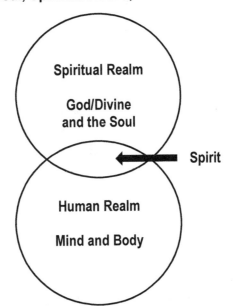

B. Guidance on Aligning with the Divine Realm

Soulful living means to live in close proximity to our souls. It means to live each day in recognition of our immortal divinity. Parker Palmer says the soul has four primary functions:[218]

1. The soul wants to keep us rooted in the *ground of our own being*, resisting the tendency of other faculties, like the intellect and ego, to uproot us from who we are.

---

[218] Palmer, Parker J. A hidden wholeness: The journey toward an undivided life. John Wiley & Sons, 2009.

2. The soul wants us to be connected to the community in which we find life, for it understands relationships help us flourish.

3. The soul wants to tell us the truth about ourselves and our world, whether that truth is easy or hard to hear.

4. The soul wants to give us life and wants us to pass that gift along, to become life-givers in a world that deals too much with death.

If we live with mindfulness of these four soul functions (or principles), we prepare ourselves to live soulfully. All four of these principles are in harmony with Lakota spirituality. The first requires us to live in alignment with the very ground of our being, which is more closely tied to emotions than thought. It is how we *"feel"* about our lives. The second is aligned relations, which is consistent with the Lakota view of *"all my relations."* Alignment with our inner truth is essential, which is the third principle. Life-giving, the fourth principle, reminds us of the importance of giving to and receiving from others. Through meditation and contemplative prayer, I attained an experienced realization that *"it's all sacred,"* much like the Lakota view. Also, I discovered that I only exist in relationship to all else in life.

> Our souls are "sacred ground." Therefore, our relationships with our souls are sacred in nature.

C. Spiritual Exercises for Forming Sacred Relationships with the Divine

How can we get in closer touch with the Divine? Here are twenty short exercises that can help you do that:

1. Cultivate gratitude, appreciation, and celebration in your day. Say thank you *"from the depths of your soul"* to family members, friends, work colleagues, caregivers, and even strangers. Show gratitude without any expectation of it being returned to you. How does your soul feel when you are thankful?

2. Walk meditatively in the woods or your neighborhood. *Witness* your thoughts and feelings. Witnessing is observing and seeing

In Sacred Relationship

without judging and labeling. Judgment prevents us from experiencing life with openness. Witnessing is an active awareness of being. How does your soul feels when you encounter life without judgment?

3. Sit in silence and consider your silent sitting to be a *prayer of listening to your soul*. What is your soul communicating to you as a feeling, thought, or image? It is common for this message to come to you later; even in a dream. Remember the soul often speaks in images.

4. Read a meaningful and fun short story to a young child you know. Imagine yourself as a child reading to this child. As a child reading to another child, we become equals. How do you imagine your soul felt about this experience?

5. Close your eyes and ask the question: *What is the most important thing my soul wants from me?* Open your eyes and write a few paragraphs about this one important thing that your soul wants most from you? Read aloud what you wrote.

6. Write a poem or short narrative about the details of a flower or stone you discover. Try to experience the flower, rock, or other physical object with all your senses, and not just how it looks. Which sensory experience was hardest for you? What does your soul want to experience and feel about the flower or stone?

7. Slow down whatever you are doing. Think slower. Read slower. Write slower. Talk slower. Slow down the preparation of your evening meal. Notice how your view of life changes when you slow down. If slowing down makes you feel impatient, describe why.

8. Linger in bed an extra ten minutes before getting up in the morning. Allow yourself to ease into your day and see the whole of life as sacred and worthy of lingering in its presence. How does this change your outlook for your day ahead?

In Sacred Relationship

9. Play a game of golf, tennis, chess, checkers, or dominoes with full awareness of yourself, especially your inner being. Did you fully enjoy the experience of playing? How did you feel if you won, or how did you feel if you lost the game? How do you imagine your soul feels about winning or losing a game? Write some short notes about the experience.

10. *Feel* music rather than simply listening to it. It helps to listen through headphones or sit alone in a quiet room. It does not matter what type of music you choose to listen to. It could be Samuel Barber's *Adagio for Strings* or Jimi Hendrix's *Purple Haze*. Describe how the music felt to you? Where do you feel the music in your body?

11. Imagine yourself as a baby looking at a butterfly for the first time. Imagine that you do not have formal words to describe the butterfly. How do your eyes, hands, and body communicate about your nonverbal butterfly experience? Describe the experience.

12. Imagine what your soul might look like. Try to imagine your soul in three separate ways. Contemplate each way you imagined your soul. Jot down your thoughts. Try to identify where your visions of the soul came from. Did they come from your church, grandmother, a movie you watched, a work of art, a song, or a book you read?

13. Imagine yourself listening to happy chirping birds on a nearby feeder. Listen with *your heart* instead of your ears. Write down what you heard and felt in listening with your heart?

14. Close the door to your room and strike up an aloud conversation with yourself about *life as a gift*. Allow the conversation to include others you trust deeply, such as your mother, grandfather, or a favorite teacher. Allow the conversation to be uncensored or unedited. What did you learn new about life as a gift? Jog down the highpoints of the conversation.

15. Sit in your kitchen and smell a couple favorite spices, like basil, thyme, or sage. Jot down on paper your olfactory experience with

In Sacred Relationship

each spice. List any memories prompted by these scents. These can be recent or distant memories.

16. Pull out a couple pictures of one or two departed loved ones or friends. Imagine the person's voice. Then imagine the person telling you that they love you unconditionally. How did you feel when you heard your loved ones say these words? Tell them you love them unconditionally. How does this make you feel?

17. Make a trip to your basement, if you have one. If no basement, select another dark and quiet room in your house. Find a safe place to sit quietly in the darkness for five minutes without closing your eyes. Imagine you are face to face with your soul in the darkness. Then turn the lights on and jot down what you learned about this encounter with your soul.

18. On your drive to work or a meeting with a friend in the morning, spend a few minutes becoming aware of one or two people in the cars sharing the road with you. Do this in a careful way so you do not get in an accident. When you arrive at work or your meeting, list a couple important questions that occurred to you about these strangers' souls.

19. Sit quietly alone in a church or other place of worship and feel the space around you. How does it make you feel? Focus a bit on your feelings of worthiness or unworthiness of sacred experience. Jot down your thoughts.

20. Sit with an elder family member or friend. Ask this person what they think about the soul. Then, ask them why they think this. Jot down your thoughts about the experience. Do this person's ideas about the soul align with or differ from your own?

Remember these exercises are simply *"prompts"* to help us connect with the Divine and our soul. Be creative and develop your own prompts. You may want to use your meditation or yoga practices to connect with your soul. Prayer is an effective way to connect with one's soul. Prayer offers an opportunity for requesting, giving or sharing, and listening. *Prayer and meditation invite us into sacred relationships.*

Reading and contemplating scriptures is a powerful way to *"personalize"* the spiritual word. The most important thing is to *"feel"* your connection with the soul, God, and the Divine. If you can, prompt yourself each day to connect with your soul. This will keep your *"soul channel"* open. It helps to keep a journal to provide a record of your experiences. In that way, you can return to them and learn from them. Try to bear in mind that the soul is ever-changing, and each time you *"step into your soul stream, it's a different place that is sometimes familiar and other times unfamiliar."*

Second Realm: Sacred Relationships with Yourself

A. Body, Mind, and Spirit

In the GSSRC, You, as a whole person, are comprised of spirit, mind, and body. It is important for us to live in sacred relationship with all three aspects of our total person, as well as living in sacred relationship with ourselves in a whole sense.

Spirit is the life force that animates us. Spirit permeates the body, but it is most connected to the heart. Spirit is an extension of the soul in the spiritual realm. The spirit is always connected to the soul, and serves as the bridge between the human and Divine realms. When a person dies, the spirit returns to its home in the soul. Within my personal spiritual worldview, the soul has many human realm embodiments. In each, a different spirit and body are given form, or manifested, by the soul. The emotions are most connected to the spirit and soul. Some call our emotions the language, or voice of the soul.

Mind consists of intelligence, cognition, memory, perception, thinking, language, and other mental faculties. It also plays an important role in both consciousness and feelings. Mind is most closely associated with the brain. Shortly, I share my understanding of consciousness, which is not entirely a function of mind.

Body is comprised of the physical body with its many systems and vital organs, and the energy body, which is how we experience both spirit and consciousness as forms of energy, or energetic experiences.

Just as there are many ways of understanding the Divine, there are many ways of understanding One's Total Self. In my own spiritual worldview, the mind, body, and spirit are inseparable aspects of our being, and they are best understood as interrelated realms of our being.

B. The Human Body

Our bodies ground us as conscious spiritual beings. *In the context of the GSSR Compass, a sacred relationship with one's body is a relationship that inspires us to see the Divine in our body.* The human body is defined in physical (material) terms and energy (non-material) terms.

Physical (Material) Body

The human physical body consists of eleven major systems:[219]

1. Circulatory/Cardiovascular: Circulates blood throughout the body
2. Respiratory: Brings air into the lungs and remove carbon dioxide
3. Skeletal: Bones that maintain the body's structure
4. Muscular: Enables movement through muscles
5. Digestive/Excretory: Absorbs nutrients and removes waste
6. Nervous/Brain: Collects, processes, transmits information
7. Endocrine: Influences body functions through hormones
8. Integumentary: Protects the body and regulates body temperature
9. Urinary/Renal: Filters the blood and eliminates body waste
10. Lymphatic/Immune: Defends the body against pathogens
11. Reproductive: Creation/birth of offspring

To promote healing and wellness of the body, it helps to envision sacred relationships with each of our body systems, as well as their interrelationships. Our bodies ground our spirits, and for that reason they are sacred ground, which we should treat with honor and respect. While all the body's organs are important, the body has five vital organs:1) Brain; 2) Heart; 3) Lungs; 4) Kidneys; and 5) Liver.

C. Sacred Relationships with Our Bodies

In Sacred Relationship with Our Whole Physical Body

The *"parts and whole"* construct is helpful in thinking about our bodies and our relationships with them. Our eleven body systems and five vital organs are integral parts of our physical body, but even taken together, these parts do not account for the "whole" of our physical body

---

[219] Shier, David, Jackie Butler, and Ricki Lewis. Hole's essentials of human anatomy & physiology. New York: McGraw-Hill Education, 2015.

because the body exists as a *totality* transcending its parts. Our whole physical body, as a functioning integrated unit, is more than the sum of its parts. In addition, the physical body exists in sacred relationship to the mind and spirit. By nature, mind, body, and spirit constitute a *Sacred Trinity"* that exists in *"Wholeness."* This is concept illustrated in Figure 9.

Figure 9: Mind-Body-Spirit Sacred Trinity of Wholeness

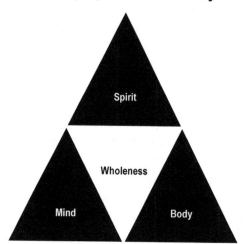

The starting point for being in sacred relationship with our body is to see our body in whole terms. This is not always easy because we are conditioned by reductionist thinking to reduce things in life, including ourselves, into pieces and parts. The danger of the piecemeal view is that it causes us to experience ourselves and our world as fractured, broken, and separated. Sometimes however, we actively strive to see ourselves in a separated way. Here is an example from my own life.

Three years ago, I was diagnosed with prostate cancer. Luckily, my cancer was detected early, and surgery "cured" me of prostate cancer. Because I was afraid my cancer might spread, I wanted to view my prostate and its cancer as separate from the rest of my body. My perspective shifted a couple weeks after my diagnosis <u>from</u> seeing my prostate as a *"non-vital"* organ that had be removed so the *"rest of my body"* could live, <u>to</u> seeing my prostate and its cancer as inseparable parts of my whole body and life. I came to realize the sacred purpose of my prostate in allowing me to give my creative life energy as a partner in conceiving my two wonderful sons Jeff and Jason, who would not be here without the role that my prostate gland played in their creation. This was a powerful spiritual realization for me, setting me on the path to invite greater wholeness into my life. Cancer and other major life traumas can serve as teachers about wholeness and other important spiritual lessons.

When we view any part of our body as separate from other body parts, we subject ourselves to disconnection and brokenness, which causes us to lose *"wholeness."* Each part of our body works in concert with all other parts to sustain us. The parts of our bodies stop working together when we fall out of balance. The same can be said about the need for the integration and connectivity of our minds, bodies, and spirits. We can overcome these disconnections and brokenness by relating to our bodies in sacred ways that cultivate wholeness. Our medical caregivers play an incredibly important role in our personal healing on the levels of body, mind, and spirit, but we still must engage in our own self-care. Medicine needs to do more to help patients play a greater role in their own healing and wellness through self-care. Self-care reminds us to live in healthy and balanced ways.

In Sacred Relationships with Our Physical Body Systems and Organs

Our body systems and organs work in concert to support and nourish our divine gift of life. Each part has a key role to play in supporting our life force energy. Within energy medicine, which is discussed shortly, each of our body systems and organs is seen as having a spiritual role to play. For now, I will simply say that:

1. The various parts of our physical body function in sacred relationship on various levels to support and sustain our whole body as an integrated unit; and

2. Our whole physical body exists because of the sacred relationships of its parts to create and maintain a living connection with our mind and spirit to fully *"enliven"* us as human beings.

Figure 10 below illustrates these connections.

Figure 10: Our Physical Body Systems' Connections and Their Connection with Mind and Spirit

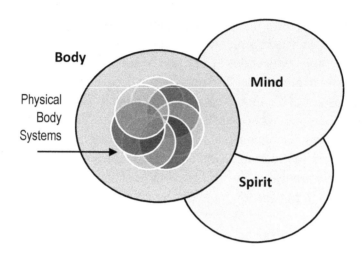

Integrated Medicine Perspective

Integrated medicine encourages us to consider the body as a total integrated system that functions through the roles played by all of its subsystems. To advance our understanding of the integrative medicine concept further, let us look at how Duke University Integrative Medicine defines integrative healthcare. *"Integrative Healthcare Is an approach to care that seeks to integrate the best of Western scientific medicine with a broader understanding of the nature of illness, healing and wellness. Easily incorporated by all medical specialties and professional disciplines, and by all health care systems, its use not only improves care for patients, it also enhances the cost-effectiveness of health care delivery for providers and payors."*[220] Duke officials go on to say that integrated medicine is: *"care that addresses the whole person, including body, mind, and spirit in the context of community.*[221] I would add that all of this works best when it is envisioned and approached in sacred relationship. Mind, body, and spirit exist in sacred relationship to one another, and they exist in sacred relationship to *"community,"* which is defined in many ways, including

---

[220] Integrative Healthcare, Duke University Medical School: https://dukeintegrativemedicine.org/leadership-program/what-is-integrative-healthcare/. Accessed January 30, 2020.
[221] Ibid.

In Sacred Relationship

family, work, neighborhood, church, school, and other social groups.

The complementary medicine and spiritual care work that my wife and I perform at Cleveland Clinic Cancer Center is a recognition that, while cancer patients primarily come to Cleveland Clinic to receive its state-of-the-art scientific medical care, they also need psychosocial, spiritual, and complementary medicine services to ensure that they are cared for as *"whole patients."* For that reason, it is important to consider our *"energy bodies,"* and the role they play in making us whole.

Interest in the connection between spirituality and health and medicine has been longstanding.[222][223][224][225][226] Moreover, a growing body of research has focused on understanding the association between spirituality and physical body systems and organs. For example, considerable research has been conducted on the brain-spiritual connection.[227][228][229][230][231] Similarly, increased research has been conducted on the relationship between the heart and spirituality.[232][233][234] More limited research has been conducted on the relationship between spirituality and the lungs, kidneys, and liver. However, considerable research has been conducted on exploring the relationship between spirituality and the immune system, especially in HIV and AIDS patients.

---

[222] Thoresen, Carl E., and Alex HS Harris. "Spirituality and health: What's the evidence and what's needed?." Annals of behavioral medicine 24.1 (2002): 3-13.

[223] Miller, William R., and Carl E. Thoresen. "Spirituality and health." (1999).

[224] Seybold, Kevin S., and Peter C. Hill. "The role of religion and spirituality in mental and physical health." Current Directions in Psychological Science 10.1 (2001): 21-24.

[225] Thoresen, Carl E. "Spirituality and health: Is there a relationship?." Journal of Health Psychology 4.3 (1999): 291-300.

[226] Seeman, Teresa E., Linda Fagan Dubin, and Melvin Seeman. "Religiosity/spirituality and health: a critical review of the evidence for biological pathways." American psychologist 58.1 (2003): 53.

[227] Devinsky, Orrin, and George Lai. "Spirituality and religion in epilepsy." Epilepsy & Behavior 12.4 (2008): 636-643.

[228] Urgesi, Cosimo, et al. "The spiritual brain: selective cortical lesions modulate human self-transcendence." Neuron 65.3 (2010): 309-319.

[229] Sayadmansour, Alireza. "Neurotheology: The relationship between brain and religion." Iranian journal of neurology 13.1 (2014): 52.

[230] Newberg, Andrew B. "The neuroscientific study of spiritual practices." Frontiers in psychology 5 (2014): 215.

[231] Mohandas, E. "Neurobiology of spirituality." Mens sana monographs 6.1 (2008): 63.

[232] Delaney, Colleen, and Cynthia Barrere. "Blessings: the influence of a spirituality-based intervention on psychospiritual outcomes in a cardiac population." Holistic Nursing Practice 22.4 (2008): 210-219.

[233] Bekelman, David B., et al. "Spiritual well-being and depression in patients with heart failure." Journal of general internal medicine 22.4 (2007): 470-477.

[234] Bekelman, David B., et al. "Symptom burden, depression, and spiritual well-being: a comparison of heart failure and advanced cancer patients." Journal of general internal medicine 24.5 (2009): 592-598.

[235] [236] [237] While much remains unexplained about spirituality and the body, we know from experience and research that our health and well-being are enhanced by spirituality and meaningful connection with the soul.

Human Energy Body

The field of energy medicine has its earliest roots in Traditional Chinese Medicine (TCM), however the use of energy in medicine dates back to 2750 BC, according to James Oschman.[238] Energy medicine (EM) is considered to be a branch of integrative medicine, which is concerned with understanding the *"human energy field (HEF),"* also known as the human biofield.[239] Energy medicine sees its role as advancing complementary and alternative medicine therapies aimed at improving human health and well-being through balancing the subtle energies of the *"hypothesized"* human energy field.[240] [241] [242]*I choose to understand the human energy field as a spiritual phenomenon, which is not measurable in Western scientific terms. In that sense, it is a "spiritual reality," and not a "scientific reality."*

In the context of energy medicine, two types of energy are recognized: 1) veritable energy fields, which can be measured; and 2) putative (subtle) energy fields, which cannot be measured with our current technology. Veritable energy fields consist of vibrational energy, such as sound, and electromagnetic forces such as visible light. The work of energy medicine practitioners involves putative energy fields, which are also known as subtle energy fields. Energy medicine is based on the belief that physical objects, including the human body, thoughts, and feelings, are expressions of veritable and putative energies. Life force energy is known as qi (Chee) in Traditional Chinese Medicine, prana in Hindu

---

[235] McCormick, Douglas P., et al. "Spiritualty and HIV disease: An integrated perspective." Journal of the Association of Nurses in AIDS Care 12.3 (2001): 58-65.

[236] Cotton, Sian, et al. "Spirituality and religion in patients with HIV/AIDS." Journal of general internal medicine 21.S5 (2006): S5-S13.

[237] Siegel, Karolynn, and Eric W. Schrimshaw. "The perceived benefits of religious and spiritual coping among older adults living with HIV/AIDS." Journal for the scientific study of religion 41.1 (2002): 91-102.

[238] Oschman, James L. Energy medicine-e-book: The scientific basis. Elsevier Health Sciences, 2015.

[239] Jain, Shamini et al. "Clinical Studies of Biofield Therapies: Summary, Methodological Challenges, and Recommendations." Global advances in health and medicine vol. 4, Suppl (2015): 58-66. doi: 10.7453/gahmj.2015.034.suppl

[240] Ross, Christina L. "Energy Medicine: Current Status and Future Perspectives." Global advances in health and medicine vol. 8 2164956119831221. 27 Feb. 2019, doi:10.1177/2164956119831221

[241] Wisneski L, Anderson L. The Scientific Basis of Integrative Medicine. Boca Raton, FL: CRC Press; 2009: 205.

[242] Human Energy Fields, The Scientific and Medical Network: https://explore.scimednet.org/index.php/2016/04/02/the-human-energy-field/. Accessed February 13, 2020.

In Sacred Relationship

philosophy, spirit in the Christianity and Native American religions, and as the doshas in Ayurvedic medicine. The three doshas (Vata, Pitta, and Kapha) are believed to be found in everyone and their particular expression defines one's specific make-up.

While much of the Western scientific community remains unconvinced of the existence of the human energy body, growing scientific attention is being given to energy medicine and the human energy field.[243] [244] [245] While scientists are in agreement about the existence of electrical fields surrounding the heart and brain, considerable reluctance exists in science and medicine to accept the concept of the human energy field or human energy body.[246] *Despite this lack of scientific acceptance, I deemed the human energy body to be worthy of inclusion in this book because it is an attempt to systematically understand the non-material aspect of human beings that plays a role in human functioning, health, and healing.*

The human energy field (HEF) is accepted widely among practitioners and researchers in the Traditional Chinese Medicine (TCM) field.[247] Tianjun Liu at Beijing University argues: *"that the human body includes both visible and invisible parts. The former is the material system and the latter is the energetic system. As matter and energy can be converted into each other, it is important to study the human body from the perspective of both the material and energetic systems, which can complement each other. The nature of the energetic system is different from that of the material system."*[248]

Reiki is an energy medicine modality that is used widely for spiritual healing and stress reduction and relaxation purposes. Reiki is a gentle touch therapy that is beneficial to patients at Cleveland Clinic. My wife Mary and I use Reiki as a stress reduction and relaxation therapy with cancer patients. In the past two years, we have performed almost 2,000 thirty-minute Reiki sessions with chemotherapy and radiation patients.

According to energy medicine practitioner and author Cyndi Dale,

---

[243] Ibid.

[244] Rubik, Beverly et al. "Biofield Science and Healing: History, Terminology, and Concepts." Global advances in health and medicine vol. 4, Suppl (2015): 8-14. doi: 10.7453/gahmj.2015.038.suppl

[245] Jain, Shamini et al. "Biofield Science and Healing: An Emerging Frontier in Medicine." Global Advances in Health and Medicine vol. 4, Suppl (2015): 5–7. doi: 10.7453/gahmj.2015.106.suppl

[246] Oschman, James L. Energy medicine-e-book: The scientific basis. Elsevier Health Sciences, 2015.

[247] Liu, Tianjun. "The scientific hypothesis of an "energy system" in the human body." Journal of Traditional Chinese Medical Sciences 5.1 (2018): 29-34.

[248] Ibid.

In Sacred Relationship

"Fundamentally, subtle energy medicine involves the study and application of the body's relationship to electric, magnetic, and electromagnetic fields, as well as light, sound, and other forms of energy. The body produces these energies and responds to these energies that are in the outer environment. Regardless of the method used, the primary purpose is to change the frequency of the body's energetic fields, channels, and centers, the three main aspects of the energetic anatomy.[249] Dale says the human energy field is very complex, but five subfields are considered most important in energy medicine:[250]

1. Auric field
2. Morphological field
3. T-fields or thought fields
4. L-fields or life fields
5. Universal light field

Auric Field

The aura has been defined variously by Western scientists and energy medicine researchers and practitioners.[251] [252] Charles Tart distinguishes four types of auras: 1) physical; 2) psychical; 3) psychological; and 4) projected.[253] The physical is explained in scientific terms as the body's bioelectric energy. The psychical and projected auras require explanation in the energy medicine and psychic research fields. Finally, the psychological aura is seen as phenomenological experience.

Prakash et. al call the aura a radiating bio-energy field surrounding the body, which is most consistent with what Tart calls the physical aura.[254] Prakash et. al. suggests: *"Since everything in this universe is made up of the same constituent particles electron, protons, neutrons etc. that means everything has an aura."*[255] Prakash et. al continue to say that human beings with a more intense aura are healthier than those with a less

---

[249] Dale, Cyndi. The subtle body: An encyclopedia of your energetic anatomy. Sounds True, 2014.
[250] Ibid.
[251] Tart, Charles T. "Concerning the scientific study of the human aura." Journal of the Society for Psychical Research 46.751 (1972): 1-21.
[252] Rubik, Beverly. "Scientific analysis of the human aura." Measuring Energy Fields State of the Science. Fair Lawn, NJ, Backbone (2004): 157-170.
[253] Tart, Charles T. "Concerning the scientific study of the human aura." Journal of the Society for Psychical Research 46.751 (1972): 1-21.
[254] Prakash, Shreya, Anindita Roy Chowdhury, and Anshu Gupta. "Monitoring the human health by measuring the biofield" aura": An overview." Int J Appl Eng Res 10.2765427658 (2015).
[255] Ibid.

intense aura, and they conclude: *"Everything and everyone in the world has a radiation field, i.e. "energy body" or aura. The human aura provides the unique signatures of the person's physical, mental emotional and spiritual state. It appears that any changes in the status of person's health be it physical, mental, emotional, or spiritual, changes his aura field dramatically and quickly. Hence complete healing of a patient will be when the distortions/deviations in his energy body disappear along with the physical symptoms of the ailment."*[256]

Despite these different definitions, findings, and assertations, Western science finds no verifiable evidence to support the existence of the auric field in psychical and projected forms. I am compelled to pose this question: *As spiritual beings, is it not possible that humans have a spiritual aura; that is an extension of the spirit beyond the boundaries of the physical body? While inexplicable in scientific terms, it remains a reality within the spiritual realm.*

Morphogenetic Field

According to Cyndi Dale, *"In biology, a morphogenetic field is a subtle field connecting a group of cells that creates specific body structures or organs. For example, a cardiac field becomes heart tissue. Morphogenetic fields (also known as morphological fields) allow an exchange between like-minded species and transfer information from one generation to another. These penetrate the aura as well as the electrical system of the body."*[257] Those associated with energy medicine and the psychic field are more likely to give credence to morphogenetic fields.

Biologist and theoretician Rupert Sheldrake envisioned morphogenetic fields (or M-fields) as invisible organizing patterns that act like energy templates to establish forms on various levels of life. One could say that Sheldrake envisioned *"spiritual DNA."* While his ideas were questioned by more conservative members of the scientific community, Sheldrake put forward his theory of formative causation and morphic fields is his 1981 book, *A New Science of Life*.[258] Sheldrake explained these fields as systems that self-organize and create structure. Sheldrake says that morphic fields are neither a type of mass nor an energy form, and yet

---

[256] Ibid.
[257] Dale, Cyndi. The subtle body: An encyclopedia of your energetic anatomy. Sounds True, 2014.
[258] Sheldrake, Rupert. A new science of life: The hypothesis of morphic resonance. Rochester, VT: Park Street Press, 1995.

they are organized by energy. Finally, Sheldrake says that morphic fields are found in cells in tissues, tissues in organs, organs in organisms, and organisms in social groups.[259]

L-Fields (Life Fields)

Harold Saxton Burr's book, *Blueprint for Immortality*, describes an invisible, intangible field of energy that he refers to as the Fields of Life, or L-fields. The L-field is a field of subtle energy that Burr equates to a jello-mold, which shapes humans into a particular form. Burr says that just as a jello-mold can be damaged, it is possible to damage the human energetic field, causing humans to lose their original shape, and become susceptible to illnesses and imbalances.[260] While many writers in energy medicine have used Burr's L-field concept, Western science struggles with the concept.

T-Fields (Thought Fields)

Dale describes T-fields as thought fields. She says, *"Each provides a blueprint and design for a different side of reality.* She adds that *T-fields also represent electrical and magnetic frequencies, the two sides of matter that combine to create the electromagnetic radiation that constantly bathes and nurtures us."*[261]

The Chakras and Meridians: Energy Medicine and Western Science

The concepts of chakras and meridians are used widely in energy medicine. According to the Chopra Center, founded by Dr. Deepak Chopra: *"The Sanskrit word chakra literally translates to wheel or disk. In yoga, meditation, and Ayurveda, this term refers to wheels of energy throughout the body. There are seven main chakras, which align the spine, starting from the base of the spine through to the crown of the head."*[262]

Meridians, or acupoints in acupuncture, are known as *"energy highways"* in the human body. In Traditional Chinese Medicine (TCM), qi (Chee) energy flows through meridians or energy highways, accessing all parts of the body. Meridians can be mapped throughout the body. TCM practitioners believe that meridians flow within the body and not on the

---

[259] Ibid.

[260] Burr, Harold S. Blueprint for Immortality: The Electric Patterns of Life. London: C.W. Daniel Co, 1991.

[261] Dale, Cyndi. The subtle body: An encyclopedia of your energetic anatomy. Sounds True, 2014.

[262] Fondin, Michelle, What Is a Chakra? Chopra Center: https://chopra.com/articles/what-is-a-chakra. Accessed February 2, 2020.

In Sacred Relationship

surface. Given acupuncture's growing receptivity in Western medicine, more scientific research has been conducted on the meridians. Researcher John Longhurst writes: *"Although additional research is warranted to investigate the role of some of the structures identified, it seems clear that the peripheral and central nervous system can now be considered to be the most rational basis for defining meridians."*[263]

Stefano Marcelli drew this overall conclusion about scientific evidence of the meridians in a 2013 article: *"Acupuncture originated in ancient China but is used around the world to treat a variety of diseases. Research has not provided conclusive evidence proving the existence of meridians, or the lines along which acupuncture needles are inserted. Currently, some members of the scientific community recognize the existence of acupuncture points but not of the meridians. Thus, when acupuncture is accepted, it is regarded as a suitable treatment for some symptoms and diseases, but its effectiveness is ascribed directly or indirectly to the nervous system."*[264]

Acupuncture, TCM, and other complementary and alternative medical practices are encompassed in the field of integrative medicine. One of the better scientific studies of integrative medicine was done by Wisneski and Anderson, who wrote: *"In order for Western medicine to have a cohesive physiological system, it must account for the existence of energy fields within as well as outside of the human body. The Scientific Basis of Integrative Medicine looks at how various forms of energy (e.g. light, sound, electromagnetism, or even prayer) translate into the chemical and electrical signals that orchestrate our physical health and mental well-being. It has been scientifically documented that factors, such as faith, prayer, and love, influence our recovery from illness. These partially understood modalities currently are spoken of in terms of 'energy,' 'healing energy' or 'subtle energy.' The Scientific Basis of Integrative Medicine borrows from the emerging field of energy medicine to present what is already known about subtle energy as well as to offer a theoretical model for the biological energy exchange that occurs with subtle energy healing."*[265]

Future research is needed to verify the existence of the chakras

---

[263] Longhurst, John C. "Defining meridians: a modern basis of understanding." Journal of Acupuncture and Meridian Studies 3.2 (2010): 67-74.

[264] Stefano Marcelli. Medical Acupuncture.Feb 2013.5-22.http://doi.org/10.1089/acu.2012.0875

[265] Wisneski, Len, and Lucy Anderson. "The Scientific Basis of Integrative Medicine." Evidence-based Complementary and Alternative Medicine vol. 2,2 (2005): 257–259. doi:10.1093/ecam/neh079

and meridians from a scientific standpoint. While we have a better scientific explanation of meridians, which was discussed earlier, it appears we have a ways to go in verifying the chakras. That said, it is important that we include a discussion of the chakras in this section of the book.

The chakras are considered to be *"focal points"* in the subtle energy body in energy medicine. They are often organized into three groups: 1) the chakras for the material or physical body; 2) those connecting the physical body and spirit; and 3) the spirit-related chakras.

Group 1: Physical or Material Body Chakras

The first three chakras, starting at the base of the spine are the chakras of the material or physical body. These include:

- **First Chakra (Root Chakra or *Muladhara*):** This is the chakra of stability, security, and other basic needs. It encompasses the first three vertebrae, the bladder, and the colon. When this chakra is open, we feel safe and courageous.
- **Second Chakra (Sacral Chakra or Swadhisthana):** This chakra is our creativity and sexual energy center. It is found above the pubic bone, below the navel, and is responsible for our creative expression.
- **Third Chakra (Solar Plexus Chakra or Manipura):** The third chakra is in the area from the navel to the breastbone. The third chakra is our source of personal power and will.

Group 2: Connection between Material Body and Spirit

- **Fourth Chakra (Heart Chakra or Anahata):** Which is found at the heart center, and functions to connect the lower chakras of material body and the upper chakras of spirit. The heart chakra is our source of love, healing, and connection.

Group 3: Spirit Chakras

- **Fifth Chakra (Throat Chakra or Vishuddha):** The fifth chakra is located in the throat area. This is the source of verbal expression and the ability to speak our highest truth. The fifth chakra includes the neck, thyroid, and parathyroid glands, jaw, mouth, and tongue.
- **Sixth Chakra (Third-Eye Chakra or Ajna):** The sixth chakra is found between the eyebrows. It is also referred to as the *"third eye"* chakra. This is our intuition and insight center.

In Sacred Relationship

- **Seventh Chakra (Crown Chakra or Sahaswara):** The seventh chakra is located at the crown of the head. This is the chakra of enlightenment and spiritual connection to our higher selves, others, and ultimately, to the divine

Figure 11: Seven Chakra Energy Centers[266]

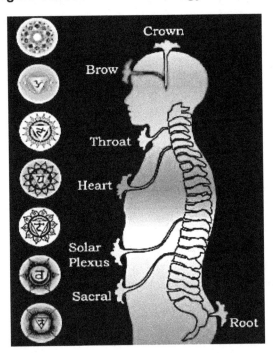

D. Exercises for Developing Sacred Relationships with the Human Body

This section presents some exercises to experience and develop sacred relationships with the physical and energy bodies. The reader is encouraged to adapt these exercises to their personal spiritual belief system. The physical body exercises should be pertinent to people of all faiths. Meanwhile, the human energy body exercises may be more applicable to those accepting the spiritual reality of the *"human energy body."* If you prefer, you can substitute your understanding of *"spirit"* for

---

[266] Understanding the Chakras, Christopher and Dana Reeves Foundation: https://www.christopherreeve.org/blog/life-after-paralysis/understanding-the-chakras-reeve-blogger-elizabeth-forst. Accessed February 4, 2020.

the *"human energy body."* Approach each exercise slowly and thoughtfully. Do not rush through the exercise.

Physical (Material) Body Exercises

1. **Body Part You Feel the Closest Relationship With**: Stand in front of a full-length mirror and use your phone camera to snap a picture of your body in the mirror. Your body can be fully clothed, partially clothed, or unclothed. Open your picture on your phone. Have a pen and notepad handy. Look at the picture of your body carefully, and then identify the <u>one</u> part of your body that you" feel" that you have the closest personal relationship with. This can be a part on the inside, such as an internal organ, or on the outside of your body, such as your face or your chest. Write down this part of the body and explain why you *"feel"* the closest to this part of the body. Try <u>not</u> to base your judgment of *"closest"* on the parts of your body that you *"like"* or *"dislike."* Now, write down in an uncensored stream of consciousness style of writing why you feel closest to this particular part of your body. Do you feel that you have a sacred relationship with this part of your physical body? Explain why you do/do not feel you have a sacred relationship with this body part.

2. **Body Part You Feel Most Distant From**: Repeat the steps in Exercise 1 to identify the part of your body you feel <u>most distant</u> from.

3. **Imagining One of Your Body Systems**: Select one major body system from the list of eleven discussed in this section. For example, use the nervous system (including the brain) as an illustration. Start by closing your eyes and imagining what the human nervous system looks like. Now, draw a picture on paper of how you imagined your nervous system to look. When finished with your drawing, copy and paste the link below in your Internet web browser and go to the website to see what the nervous system looks like: <u>http://anatomysystem.com/nervous-system/</u>. How did your imagined view and the actual picture of the human nervous system compare? What did you learn from this exercise?

In Sacred Relationship

4. **Whole Physical Body Sacred Relationship**: Use your body photograph in this exercise. Study the picture carefully, and then close your eyes for a minute and envision your whole physical body as best you can. Hold on to your whole-physical body vision in your mind. Now, open your eyes and write down two things you consider to be most sacred about your whole body, and two things you feel least sacred about your whole physical body. Read what you have written. Now, write down your main reasons for the two things you listed as most and least sacred about your whole physical body. What did you learn from this exercise?

Exercises to Experience Your Energy Body

1. **Sacred Relationships with the Seven Chakras:** Examine the chakra map in Figure 11 above and re-read the short descriptions of the seven chakras. Both appear earlier in this section. List the seven chakras in this order on a sheet of paper:

   - Root (stability, security, basic needs)
   - Sacral (creativity, sexual energy)
   - Solar Plexus (personal power and will)
   - Heart (love, healing, connection)
   - Throat (speaking personal truth, self-expression)
   - Brow (insight, intuition, psychic power)
   - Crown (enlightenment, connection to divine)

   For each of the seven chakras, visualize its location in your energy body and rate whether you feel your sacred relationship with each chakra is strong, medium, or weak. Use Table 1 below as a guide. It may be helpful for you to place your hands on each of the chakra locations. Physical touch helps us *"tune into"* our physical and energy bodies. Therefore, in part, Reiki works as a stress reduction and relaxation technique.[267]

   As described earlier, Reiki is a gentle touch therapy used to promote energy balancing and healing in the body. In my Reiki practice, I use 12 hand positions that include the seven chakras,

---

[267] Learn more about Reiki therapy here: https://www.reiki.org/faqs/what-reiki

as well as other energy locations on the front, sides, and back of the body. My Reiki sessions start with the head area, then move downward to the neck and chest, then the midsection of the body, and finally to the lower body (legs and feet) and the back. Reiki is one of many energy medicine practices used to strengthen the human energy body. I have found in my work with cancer patients that even if patients do not believe in the concept of the human energy body, they receive significant emotional and physical comfort and relief from Reiki therapy. Almost 40 percent of patients treated by my wife and me go to sleep during a Reiki session, which is a very good sign of the comfort and relaxation benefits they receive from Reiki.

Table 1: Assessing Your Sacred Relationships with Your Seven Energy Chakras

Chakras	Strong Sacred Relationship	Medium Sacred Relationship	Weak Sacred Relationship
Root			
Sacral			
Solar Plexus			
Heart			
Throat			
Brow			
Crown			

2. **Improving Sacred Relationships with Your Chakras**: Review your work in Exercise #1 above. For each chakra, list in your notebook one important thing you could do to improve your sacred relationship with each chakra. Try to identify specific improvements, actions, or changes. When done, spend some time contemplating these chakra sacred relationship improvements.

Identify how you will undertake one or two of these improvements in the near future. For example, you may decide that you need to improve your knowledge of the chakra system. Here is an example of an action to improve your sacred relationship with your Root Chakra. The suggested improvement is to spend less time worrying about your personal financial resources and needs, and instead work with a personal financial advisor to develop a plan to increase your personal wealth gradually over the next three years. Money and financial security tend to be associated with the Root Chakra.

3. **Exploring Your Body Energy**: In this exercise, you will have an opportunity to explore how your body emotional energy *feels*. Our emotions provide us with valuable insights about what is going on in our energy bodies. Select one day in your week and set aside 5 minutes three times during that day when you can explore how your body energy feels. One approach is to do this first thing in the morning, at midday, and before going to bed. Sit quietly and tune into your feelings. Identify your dominant feeling and make note of it in your notepad or journal. Some examples of dominant emotions are anger, sadness, fear, anxiety, depression, frustration, being overwhelmed, confusion, and resentfulness.

Try to tune into the part of your physical body where you feel this dominant emotion. For example, is your breath constricted, and do you feel tension in your chest? You may discover you have *"mixed emotions,"* which means you are experiencing a combination of emotions. After your third session, review what you learned about your body emotional energy. This is an exercise you could do daily if it helps you to tune into your body emotional energy. You can change your emotional energy through intention and action. Closing your eyes and taking several relaxing breaths can relieve chest tension. If you feel tension in your face, focus your attention in these areas and relax the muscles in this area.

E. Forming Sacred Relationships with Your Mind and Spirit within Consciousness

Consciousness: A Convergence of Mind, Body, and Spirit

The human experience of life is an integrated one, making it very difficult to distinguish the influences of spirit and mind on our experiences. We tend to have a better handle on the influence of the body on our experiences, but at times that is hard to identify. *In Sacred Relationship* encourages us to envision ourselves as whole or complete. The GSSR Compass helps us view mind, body, and spirit as dynamic forces that converge and give rise to consciousness. While there is little consensus among experts on the definition of consciousness, *I define consciousness as our subjective inner sense of ourselves and others. Within consciousness, we experience "self-consciousness" and "other-consciousness," as well we sense that our inner and outer worlds exist as separate and connected realities. Consciousness is the resulting field of relational awareness and understanding that emerges through the convergence of mind, body, and spirit.* This emergent view of consciousness provides a plausible conceptual understanding of how mind, body, and spirit figure into consciousness and the human experience. Figure 12 visualizes the relationships just described.

Figure 12: Mind, Body, and Spirit and Consciousness©

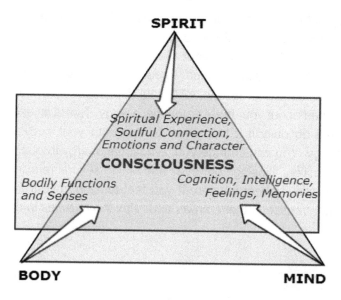

In Sacred Relationship

I would like to share a little bit more about my spiritual worldview. It is the Soul, our most fundamental spiritual essence, that works through Spirit, to give birth to *"life force"* in the Mind and Body. The role of Spirit, as an extension of Soul, is to *"animate"* our beings. Once again, the Soul is the locus of our natural divinity, which makes it closest to God.

In Sacred Relationship encourages us to view Soul, Spirit, Mind, Body, and Consciousness as interrelated dynamic processes. The experience of any of these realms as fixed states is a trick of memory and our mental conditioning (habits of thought). Memories are vital inputs to cognitive processing, including learning. In short, we remember (record) past experiences, store them away, and retrieve them for use in current and future life experiences. The problem is that our memories are *"faded pictures"* of our earlier experiences with a thing, person, or event. Our memories change over time. While we may remember the essential aspects of past experiences, we forget aspects, and we fill in the blanks with new information. In short, the stories we tell about our memories change over the time.

I wanted to share an additional observation about how we think about and experience our consciousness. Two decades ago, I stumbled upon Promethea, a comic series by Alan Moore. I was entertained and intrigued by Moore's observation about consciousness: *"Consciousness, unprovable by scientific standards, is forever, then, the impossible phantom in the predictable biologic machine, and your every thought a genuine supernatural event. Your every thought is a ghost, dancing."*[268] The beauty of Moore's observation is that consciousness is ghost-like, and therefore appears to be supernatural in our experience of it. There is nothing fixed or solid about our thoughts, or even feelings for that matter. In this sense, thoughts and feelings possess a phantom spiritual quality.

Advice on Connecting with Mind and Spirit within Consciousness

To prepare you for the exercises for forming sacred relationships with mind and spirit, I wanted to share some working definitions of the primary attributes associated with mind, body and spirit, which are identified in Figure 12 above.

---

[268] Moore, Alan, Promethea Book Five. DC, 2006.

Spirit within Consciousness

Spirit is most closely connected to Soul, and therefore Spirit is our source within consciousness for soulful connections, spiritual experiences, character, and emotions. Each of these attributes has central importance as *"spiritual contributions"* to consciousness, which is once again an emergent outgrowth of mind, body, and spirit coming together. Let us consider these definitions:

- *Soulful connection and grounding* are how the Soul manifests in our lives through spirit. These connections ensure that we are never separated from our souls. Spirit's job is to keep our *"soul channel"* open, ensuring that our lives serve the needs of our soul.

- *Spiritual experiences* are *"experiential evidence"* of the Spirit's connection with the Soul. The sense of spiritual presence, or a numinous experience, informs us of the Soul's presence, and its urgings and longings. Spiritual experiences are deeply moving in an emotional sense, and they feel very personal to us. Through spiritual experiences, the spirit makes the soul's desires known.

- *Emotions* are *"a complex reaction patterns, involving experiential, behavioral, and physiological elements, by which an individual attempts to deal with a personally significant matter or event. The specific quality of the emotion (e.g., fear, shame) is determined by the specific significance of the event."*[269] Emotions are tied to the Spirit and Soul rather than the Body or Mind.

- *Character* is more fluid than what psychologists call *"personality,"* and character is highly responsive to context, or what is happening in our lives at any particular time. Personality is seen as the qualities that make us who we are, while character consists of the traits we embody and live by. Moral compass and ethics are connected closely to character.

Mind within Consciousness

Mind consists of intelligence and cognition, memory, perception, thinking, language, and other mental faculties. It also plays an important

---

[269] Emotion, APA Dictionary of Psychology, https://dictionary.apa.org/emotion. Accessed February 29, 2020.

role in both consciousness and feelings. Mind is most closely associated with the functioning of the physical brain. Let us examine some definitions of these primary faculties of the mind:

- *Cognition and Intelligence* consist of all forms of knowing and awareness, such as perceiving, conceiving, remembering, reasoning, judging, imagining, and problem solving. Along with affect and conation, it is one of the three traditionally identified components of mind. Intelligence is often seen as a partial measure of cognitive ability, such as Intelligent Quotient (IQ).[270]

- *Memory* is the ability to retain information or a representation of past experience, based on the mental processes of learning or encoding, retention across some interval of time, and retrieval or reactivation of the memory.[271]

- *Perception* is the process or result of becoming aware of objects, relationships, and events by means of the senses, which includes such activities as recognizing, observing, and discriminating. These activities enable organisms to organize and interpret the stimuli received into meaningful knowledge and to act in a coordinated manner.[272]

- *Thinking* is cognitive behavior in which ideas, images, mental representations, or other hypothetical elements of thought are experienced or manipulated.[273]

- *Language* is a system for expressing or communicating thoughts and feelings through speech sounds, written symbols, or nonverbal communication.[274]

Body within Consciousness

The body participates in consciousness through its body functions, such as respiration and digestion, and its senses, such as smell

---

[270] Cognition, ibid.
[271] Memory, ibid
[272] Perception, ibid
[273] Thinking, ibid
[274] Language, ibid

and touch. Because an earlier section on forming sacred relationships with physical and energy bodies dealt with these issues, they are not included here.

Exercises to Experience Mind and Spirit within Consciousness

I encourage readers to approach the exercises for experiencing mind and spirit within consciousness in a holistic and integrated manner. In other words, try not to separate mind and spirit in the exercises. Instead, focus your attention on what is going on in consciousness in an integrated sense. For simplicity sake, it may be helpful to view consciousness as a constantly changing *"container,"* in which our thoughts, feelings, and experiences arise and change. Imagine observing your thoughts, feelings, and experiences as actors in a movie you watch on a television screen. This will make these exercises more accessible and meaningful.

Here is an exercise to help you form sacred relationships with your mind and spirit within consciousness:

Exercise: Experiencing a Class Lecture at Sorbonne University:

In preparation for the exercise, visit the Sorbonne University website and spend five minutes or so reviewing information on its website. The University's website link is found in the footnote below.[275]

Now, close your eyes. Take three or four easy breaths to relax your body. Take a moment to set an intention to relate to this exercise in a sacred way, which means you wish to encounter your Mind, Body, and Spirit within Consciousness in this exercise with reverence, respect, trust, openness, and a willingness to learn. *The key to this exercise is envisioning this experience as "sacred," which means that spirit and soul are actors in the experience. See your consciousness field as "sacred ground" and everything you experience as being sacred in nature.*

Image yourself walking around the Sorbonne University campus in Paris, France. You see many students, magnificent buildings, green lawns, and beautiful gardens. You hear people speaking in French. You are struck by the beauty of the French language. You approach your walk with reverence and gratitude for this amazing university. Your Body carries you in the direction of the Digital and Cognitive Sciences Building because you see a fascinating modern art sculpture in the yard outside the building.

---

[275] Sorbonne University website: http://www.sorbonne-universite.fr/en/university

Your Mind is curious about the meaning of the sculpture, and your Spirit is lifted by its beauty and elegance. You enter the building, and read the kiosk, which informs you that a special lecture is being held now about "*Art and Artificial Intelligence.*" You decide to stop in the classroom and sample the lecture. You remember the modern art sculpture in front of the building and think that it seems to be about the connection between art and digital and cognitive sciences.

You take a seat in the back of the classroom, which is filled with fifty or more students. You are mesmerized by the lecture, which is about art as an inspiration for the development of artificial intelligence technology. This is a topic you have never thought about before. You allow yourself to explore the lecture with your imagination. You think of a question you would like to ask the professor about an idea presented in the lecture. When the time comes, you raise your hand, the professor acknowledges you, and you carefully ask your question. You ask: "*Professor, you have been talking about how art is an inspiration for artificial intelligence. Is it also possible that artificial intelligence is an inspiration for art?*"

The professor ponders your question for a few seconds, and replies to you with a question: "*Would you like me to answer the question from the standpoint of an artist or artificial intelligent agent (AI agent)?*" The class erupts in laughter. Your heart leaps because the whole meaning of your question changes in light of the professor's question. At this point, your Body is responding to the experience with a quickened heartbeat and for a second your breath is taken away. Then, you find yourself in the ensuing seconds scouring your inner world (your field of consciousness) for guidance on how to respond to the professor's question. You experience a slight feeling of panic, but you are not certain if this feeling is mental or emotional. You feel a bit vulnerable because your sense of reality has been challenged. The professor's question took you off guard. You do not want to appear uninformed in responding to the professor's question, but you know you must respond quickly because all eyes in the classroom are on you. You take a deep breath, and suddenly an answer pops into your head, "*I am neither an artist nor an artificial intelligence agent, and therefore I am not sure whether I would really understand your answer from either perspective. Could you offer an answer from the standpoint of the artist and the artificial intelligent agent?*"

The professor replies, *"The artist says I am inspired by anything that is new, and often things I know nothing about because they cause me to reach inside myself for a novel artistic representation. I know nothing about artificial intelligence, and therefore it could inspire my art. The artificial intelligence agent replies: I have never thought about myself as a source of artistic inspiration because I was not programmed to do this. I am programmed to interact with you, and if that inspires you in an artistic way, then perhaps I can play some role in artistic inspiration."*

The professor's reply stirs many thoughts and feelings inside you. You thank the professor, and you share that you learned that both the artist and the artificial intelligence agent have a similar lesson to teach each other, which is that the interaction with the unknown and new can bring about artist inspiration if we are open, honest, and willing to learn. Moreover, artistic inspiration is something we ultimately find inside ourselves where Mind and Spirit arise and change within our consciousness. The professor smiles and says: *"Thank you for taking your journey inward and finding your own artistic inspiration."*

Now, take a few moments to think about this exercise and what you personally learned about being in sacred relationship with your Mind and Spirit within Consciousness. When viewed as "sacred," now note how you experienced "curiosity" in your consciousness. How does curiosity make you feel, and what does it make you do? Pay attention to how you experienced "surprise" in your consciousness. Did it make you feel afraid or excited in a positive sense? Note how you experience your own worldview, or overall way of seeing the world, while listening to the lecture on art and artificial intelligence in the context of the classroom in Paris. Did you feel clever and intelligent in your exchange with the professor, or did feel vulnerable and confused? How did you visualize the event? Did you see it as a movie, or a series of photograph images in your mind?

Realm 3: Forming Sacred Relationships with Other People

Relating in a Sacred Way with Others

This section is a guide to experiencing our relationships with other people as sacred. Many of us are <u>unaware</u> of the *"potential"* for sacredness in our personal relationships, until something happens awakening us to this reality. What can awaken us? Sometimes circumstances that threaten the stability and longevity of an important personal relationship bring about this awakening. Serious martial problems, life-threatening illnesses, major financial disasters, and other notable events can shock us into an awareness of the sacredness of a relationship. Crisis can either spark our higher or lower consciousness. *This book is an invitation to actively cultivate awareness of the "higher sacred potential" of our relationships.*

Four Types of Core Sacred Relationships

There are four core sacred relationship types. These types include human-to-human, as well as virtual relationships mediated by digital technology. These four types are shown in Figure 13 below. Sacred relationships can exist in each of these relationship types. The first, and perhaps most common, is the *two-party relationship*, which could take the form of a marriage or other committed relationship, friendships, two-party work relationships (partnerships), sports or play relationships, mentor-mentee relationships, teacher-student relationships, doctor-patient relationships, and a host of other two-party relationships.

The *family web relationship* most often involves more than two people. It can be a triad (mother, father, child), a nuclear family web (parents and the children), or an extended family web (nuclear family plus other relatives). With the growing number of marital divorces and separations, and an increase in alternative parenting arrangements, family web sacred relationships have increased in complexity. The *work web relationship* consists of several people working together for a shared work purpose. These relationships can be peer-based or hierarchical. They usually involved many people. Finally, there is the *community web relationship*, which involves multiple parties in a neighborhood, tribal culture, membership organization or association, church or spiritual community, hospital or other healthcare organization, school or educational community, military unit, musical group, political party, and other types of shared interest organizations and associations. The relationship becomes more complex with more participants involved.

While it may be easier to associate sacred relationships with two-party and relationships involving a small number of people, sacred relationships also are possible in the other three types of core sacred relationships. The key is to find spiritual purpose and soulful connection in any type of relationship. Religious and spiritual organizations often see these purposes more easily because these organizations focus on spiritual purposes. Any societal institution, including education, medicine, government, and business, can be guided by a spiritual purpose defined in a wide variety of ways. We should aim to cultivate sacred relationships in all life arenas.

Figure 13: Core Sacred Relationship Types

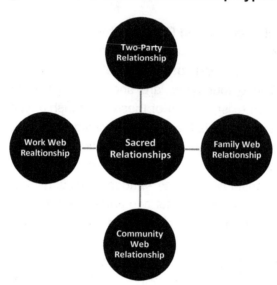

People with Whom We May Relate in a Sacred Way

Our lives unfold through the relationships we form with others. It is extremely hard to find an aspect of our lives that does not involve relationships. Each of our lives is a dynamic *"web of relationships."* During our lives, we form and grow relationships with:

- Parents and grandparents
- Spouses, life partners and companions
- Children, grandchildren, and great-grandchildren
- Extended family members
- Friends on various levels of closeness
- Work colleagues and associates
- School classmates and teachers
- Religious and spiritual community members
- Those we serve with in the military
- Caregivers of several types, including spiritual caregivers

- Coaches, mentors, role models, and other advisors
- Neighbors and people within our larger community
- Teammates and playmates
- Pets with whom we share our lives
- Our ancestors and deceased friends and loved ones

Indicators of Sacred Relationships

How does one know if a relationship is sacred in its meaning and purpose? Often, we do not know until the relationship unfolds. Although in some cases, the parties involved know the relationship is sacred from the onset. Sacred relationships, like all relationships, lack perfection, and they are filled with learning and growth. Sacred relationships must be grown into like all other relationships. They need time and nurturing. Here are five important indications that a relationship has sacred potential:

1. **A Sense Exists that the Relationship is Very Special:** Sacred relationships have spiritual purposes that are often *"felt"* deeply before they are recognized consciously. These purposes are often transformational.

2. **Unconditional Love or the Search for It Exists**: Spiritual purpose is closely tied to unconditional love. In short, the parties involved in sacred relationships commit themselves to loving each other for who they really are. This duty to authenticity is a sign that the relationship is sacred.

3. **Shared Feeling of Deep Connection:** Sacred relationships are characterized by deep sharing, and transformation. Always, strong emotion is found in sacred relationships. The involved parties share these feelings.

4. **The Relationship is Multifaceted and Evolving**: Sacred relationships are rarely about one or two things. Usually they grow in various directions and on various levels over time. Soul connections are powerful and lingering.

5. **A Sense of Sacred Oneness Exists**: Sacred relationships are characterized by oneness and unity. *"We"* and *"us"* comes first.

The parties in the relationship *"sense"* their relationship is sacred and deeply important.

Look for these signs of sacred potential in your relationships. Encourage these qualities to shape your relationships. If you are like me, you will be pleasantly surprised by the power found in sacred relationships. Their power can enable and propel us.

Sacred Relationship Toxicity

All types of relationships, including sacred relationships, can be undermined by *toxicity*, which includes our relationships with our spouses and life partners, parents, siblings, close friends, and other people. *By definition, a toxic relationship is one that is dangerously unhealthy and potentially poisonous in emotional and spiritual ways to those involved. Toxic sacred relationships cause great pain and suffering because they afflict the most important relationships in our lives.* What are the most common signs of toxic sacred relationships? The top five indicators are:

1. **Lack of Authenticity**: You feel you cannot be who you really are in the relationship. You must hide who you are, and what your soul is asking of you in life.

2. **Nobody Feels Fulfilled**: You feel you cannot do anything right in the relationship. Everything is a source of confusion, conflict, and pain because the focus of the relationship is not on meeting the parties' soul purposes.

3. **Vicious Cycle of Negative Emotions**: Those involved have no room for growth and development. This means the parties are *"locked"* into a vicious cycle of anger, hurt, fear, resentment, unforgiveness, and other deep negative emotions.

4. **Lack of Healthy Exchange**: The relationship is all about the other person and never about you. The relationship lacks the *"give and take"* necessary to fulfill soul purposes.

5. **Absence of Depth and Meaning**: The relationship is superficial and lacks meaning and purpose. Anything short of soulful meaning and purpose leads to superficiality.

Can toxic sacred relationships be repaired or healed? The answer is that it depends on the depth of the underlying problems, and the participants' willingness and ability to change. Again, sacred relationships have soul purposes. If these soul purposes remain unknown and unfulfilled, then it is highly unlikely the sacred relationship can be *"detoxed."* Sacred relationships require deep healing by the individuals involved. Often, people do not know how to approach deep relationship healing. Alternatively, the relationship participants concentrate on making easy *"surface"* changes

> We must take part in our relationships as though they are "sacred ground," which in fact they are!

that have no impact on sacred relationship toxicity. The starting point is always the same: See life in sacred terms; all of life, including the people who take part in our lives. The second step is to act in line with this sacred precept. *We must participate in our relationships as though they are "sacred ground," which in fact they are!*

As I look back on my life, I had little guidance from others on how to form healthy sacred relationships. I made many mistakes in the relationships I was involved in. The important thing is that we learn, change, and grow. I had no idea such a thing as a sacred relationship even existed. I unconsciously internalized the qualities and dynamics of the relationships of my parents, aunts and uncles, neighbors, and friends. I either had very few sacred relationship role models available to me, or they were there, and I just did not recognize them.

In my early years, my best sacred relationship role models were my grandmothers, who showed me unconditional love, and encouraged me to see the bigger picture in life. My grandmothers helped me see the importance of perennial wisdom; that is wisdom we inherit from our ancestors, and wisdom that has meaning over time. I saw people using other people and getting used by others. It was all about getting what everyone wanted. From that exposure, I saw relationships as avenues to get what I wanted in life. They were utilitarian relationships dominated by manipulation, ignorance, and selfishness. In short, they were *false ego-*dominated relationships.

In today's world, I find the whole notion of *personal branding* to be deeply troubling because of its inauthenticity and propagation of self-serving or narcissistic relationships. Personal branding is an outgrowth of consumer culture, which encourages people to *"sell themselves"* to others to get what they want in life. Personal branding is the antithesis of sacred relationships. In my own life purpose coaching work, I have met people who simply wanted me to write a personal branding narrative they could *wear like a new suit of clothing.* I discovered that I have never been able to really help these clients. What is the alternative? Be who you really are, and work at being the best at that every day of your life.

Pathways to Sacred Relationships

Opportunities exist for us to grow sacred relationships with several types of people in our lives. Usually, our greatest opportunities are with those with whom we have the closest relationships. *Rather than thinking of our close relationships as opening the door to the sacred, I believe it is more correct to see our sacred relationships giving rise to closeness.* Sacredness requires closeness, and sacredness comes about when souls touch each other in deep and meaningful ways. This raises the question: *What is needed to form sacred relationships?* Twelve pathways to sacred relationship-building are offered here. Please try not to see these as a set of competencies you must acquire quickly. Learn each by working with them one at a time. The twelve pathways are to:

1. Bring your soul in close proximity to another so they can touch;
2. Recognize that all lives matter because life itself is sacred;
3. See and honor your own wholeness and that of others;
4. Respect people in their own right, and not because of what we expect from them;
5. Perceive people as spiritual beings with souls;
6. Feel gratitude and appreciation for others;
7. Form meaningful relationships that involve giving and receiving;
8. Relate to others authentically with trust and openness;
9. Demonstrate compassion and empathy for others;
10. Grow and adapt our relationships with others;
11. Relate positively, peacefully, and joyfully with others; and
12. Forgive ourselves, forgive others and seek others' forgiveness.

As I said earlier, not all our relationships are seen on the surface as sacred in nature, yet they are, because each of us is a spiritual being and our souls touch each time we relate to another person. I believe even

> Rather than thinking of our close relationships as opening the door to the sacred, I believe it is more correct to see our sacred relationships as giving rise to closeness.

our *"virtual"* relationships, or those meditated by digital technology and the Internet, are sacred. This idea is developed in my new research on the influence of digital culture and technology on religion and spirituality. I encourage the reader to look at each person in their lives as offering them something of sacred importance. The twelve pathways are designed to overcome spiritual obstacles that stand in the way of building and growing sacred relationships. Spiritual stumbling blocks can include: personal insecurities; selfishness; lack of personal and interpersonal insight; conditioned ways of relating to others we learned from our parents or other people with an early life influence on us; things that trouble us about another person that are negative traits we possess ourselves; lack of trust; lack of respect; unwillingness to let go of fear or anger; and the inability to be with others in an open and authentic way. Identify your personal spiritual stumbling blocks by asking this question: *"What is preventing me from relating with others in a sacred way?"*

Exercises for Using the Twelve Pathways to Sacred Relationships

1. **Create Soul Proximity:** Our distance with people is often an *"imagined"* distance that stems from our assumptions about ourselves and others. Select someone close to you, and then identify four things that are spiritual stumbling blocks to growing your sacred relationship with them. The first two should be spiritual obstacles you bring to the relationship, and the second two are spiritual obstacles the other person brings to the relationship. List the four, identify what you believe causes them, and then identify how these stumbling blocks can be eliminated or at least lessened in their impact. Do not overanalyze the relationship with respect to spiritual obstacles. Pick the first spiritual obstacle that you bring to the relationship and work that through. *As you work through this first obstacle, hold an intention to be equally fair to yourself and the other person.* This is not just

In Sacred Relationship

about you getting what you want, or to justify your attitudes or behavior. Instead, it is about you and the other person getting what you really need in a shared spiritual way. As a last step, visualize your spiritual relationship as closer absent these obstacles. When you are ready, re-engage this person with the awareness that you have removed your two spiritual obstacles. Watch over time whether your relationship grows closer and more spiritually significant.

2. **All Lives Matter and They are Sacred:** Open your journal or notebook and list two people you can easily accept 100 percent and two people you find it almost impossible to accept at all. These can be people you know personally or those you do not know personally, but you have read or heard about. It could be the President of the United States or a powerful character in a movie you just watched. Now describe the difference between these two sets of people. What makes it easier for you to accept the first two people and impossible to accept the second two people? Do you consider the four peoples' lives to matter equally and see all four as sacred?

3. **Promote Wholeness:** What does wholeness mean to you? I will give you a hint. Wholeness includes everything in our lives, including our toxic relationships, dysfunctional families, emotional traumas, illnesses, financial setbacks, shortcomings, prejudices, and anything else. Describe what it is like for you to be whole. What makes you feel whole? Write down your thoughts on wholeness. What role do you see others playing in your wholeness? What role do you play in helping others become whole?

4. **Respect People in Their Own Right and Not for Your Sake:** What does this principle mean to you? Identify a couple people you believe you need to work harder at accepting in their own right and not because of what they can do for you. Write down your ideas about how you can do a better job of respecting these people in their own right.

5. **Perceive People as Spiritual Beings with Souls:** What does this mean to you? In what ways is it easy for you to see people in this way, and what makes this difficult? Select one person who is especially important to you, and then describe how you see them spiritually. What do you see as the most important *"soul issue"* that influences their life? This can be something good or something not so good. Finally, how can you help this person with this soul issue? What are your soul struggles, and how can others help you with them?

6. **Feel Gratitude and Appreciation for Others:** Select one person you know well and find out why you feel grateful for them, and why you appreciate the spiritual role they play in your life. Next, describe whether you experience this person as being grateful for you and what spiritual role you play in their life.

7. **Meaningful Giving and Receiving Relationships:** List three spiritually meaningful things you regularly give to others and why. Then, list three spiritually meaningful things others regularly give you and why. Does this giving and receiving occur in the same relationships or different ones?

8. **Relate to Others Authentically with Trust and Openness:** What does authenticity mean to you? How are trust and openness aspects of authenticity? Identify someone you are almost always authentic with, and one person you are almost never authentic with? Find the main reasons in each case. What motivates you to be authentic with others? What motivates you to not be authentic with others?

9. **Demonstrate Compassion and Empathy for Others:** What do the qualities of compassion and empathy mean to you? Identify a couple situations in your life that have helped you become more compassionate. Do the same with your ability to be empathetic. Which is easier for you, compassion, or empathy, and why?

10. **Grow and Adapt Our Relationships with Others:** On a scale of 1 to 10, with 1 being the easiest and 10 being the hardest, rate

your <u>overall</u> ability to grow and adapt relationships in two areas of your life: a) in your family; and b) in your work or career. What is the number one obstacle making it hard to grow and adapt your family relationships? Do the same for your work or career relationships.

11. **Relate Positively, Peacefully, and Joyfully with Others:** Review your closest relationships, and then identify those relationships in which you experience the greatest positivity, joy and peace. Then, identify those relationships in which you feel the least positivity, joy, and peace. What accounts for the difference between the two sets of people? Identify a couple things you can easily do to increase the positivity, joy, and peace in your relationships.

12. **Forgive Ourselves and Others and Seek Others' Forgiveness:** Identify three close relationships in which you feel the greatest ability to forgive or be forgiven. Then, identify three close relationships where forgiveness is the hardest. What makes the greatest difference in the two? What does forgiveness provide to you and others? Why is it hard at times to forgive others and seek others' forgiveness?

Realm 4: Sacred Relationships with the Whole of Life

The Whole of Life can be understood as all life on the Planet Earth, which includes everybody and everything found everywhere. How we relate to the earth has a bearing on whether we live in harmony and balance. We will approach this fourth realm in the context of Planet Earth.

About Planet Earth

Earth is estimated to be 4.5 billion years old. It is the third planet from the Sun, and it is the only planet known to support life as we know it. The Earth's surface area is estimated at 197 million square miles. The Earth's orbit around the Sun takes just over 365 days. Earth's expected long-term future is closely linked to the Sun. Scientists expect that over the next one billion years, solar luminosity will increase by 10 percent, and over the next 3.5 billion years, the Sun's luminosity is expected to increase by 40 percent.[276] Is global climate change a reflection of this long journey to much hotter Planet Earth? This may be an important reality to ponder.

The Earth's total population is estimated to be 7.7 billion people. China is the most populous country with a population of just over 1.4 billion, and India has a population of over 1.355 billion people. By the year 2030, India is expected to become the most populous country in the world, while China is projected to see a loss in population.[277] With the future growth of the world's population, additional demands will be placed on our natural environment to support this growth. With this growth in mind, we must learn to live more sustainably; a reality that the consumer economy so far has failed to recognize in an appreciable way.

Top environmental threats facing Planet Earth are global climate change, pollution, deforestation, reduced biodiversity, and the melting polar icecaps.[278] About 60 percent of Americans see global climate change as a major environmental threat, which is much lower than threat ratings by citizens in other countries such as France (83 percent), Japan (75 percent), Germany (71 percent), and Canada and the UK (66 percent).[279] Perhaps the grounded spirituality and sacred relationship compass presented in this book can help us see the sacred nature of our natural resources.

An Approach to Sacred Relationships with the Earth

The Earth consists of an extraordinarily complex web of life, which is comprised of two interconnected living systems: "the *natural environment system;*" and "the *human system.*" A variety of subsystems are found within each of these two major systems.

Principles for Envisioning the Earth as Sacred

Here are five principles that can help that help us relate to whole of life on Earth in a sacred way:

1. Humans and nature co-create the world in which we live.

---

[276] Sackmann, I.-J.; Boothroyd, A. I.; Kraemer, K. E. (1993). "Our Sun. III. Present and Future". Astrophysical Journal. 418: 457–68.

[277] World Population Review: http://worldpopulationreview.com/. Accessed February 7, 2020.

[278] U.N. Environment Report, 2018: https://www.unenvironment.org/annualreport/2018/index.php. Accessed January 26, 2020.

[279] Fagan, M., and Huang, C., A look at how people around the world view climate change, Pew Research Center: https://www.pewresearch.org/fact-tank/2019/04/18/a-look-at-how-people-around-the-world-view-climate-change/

2. Humans exist in dynamic relationship with nature. Humans change, and their relationship with nature changes.
3. Humans should strive to live in harmony and balance with nature; not at its expense. We must work at this balance all the time.
4. Nature is everchanging, and its systems must reestablish balance with one another constantly.
5. Human life is sacred because of its intimate "*spiritual*" connections with nature, and nature is sacred because it is animated by living spirit.

Figure 14: Planet Earth's Web of Life Systems

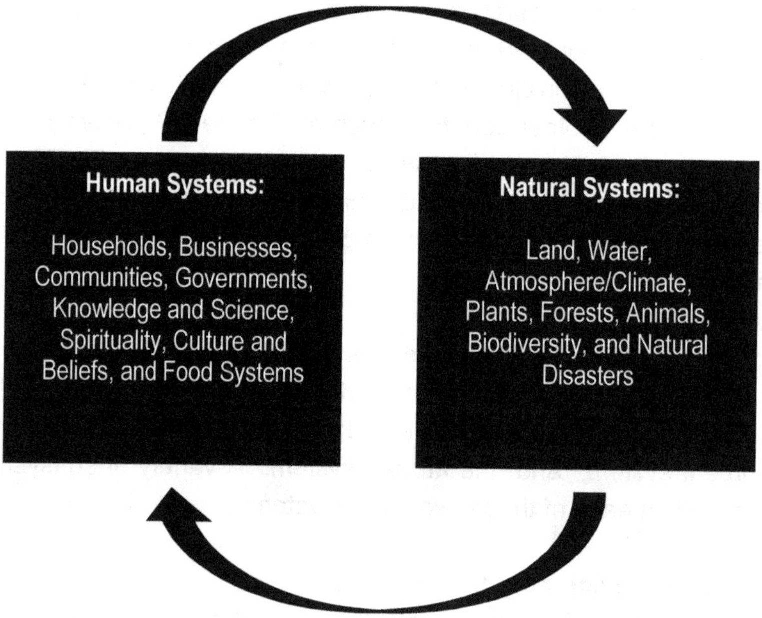

Human Systems:

Households, Businesses, Communities, Governments, Knowledge and Science, Spirituality, Culture and Beliefs, and Food Systems

Natural Systems:

Land, Water, Atmosphere/Climate, Plants, Forests, Animals, Biodiversity, and Natural Disasters

Exercises for Strengthening Our Sacred Relationship with Earth

The exercises in this section help us to strengthen our scared relationships with earth on three interrelated levels:

1. Planet Earth as a whole;
2. Endangered wildlife species; and
3. Sacred lands and sites.

In Sacred Relationship

Planet Earth as a Whole Exercise

In your notepad or journal, list three *"unique"* ways that you view the whole of Planet Earth in a sacred relationship sense. Here are a few examples of how one might view the whole of Planet Earth:

- As a constantly changing green and blue sphere seen from a space station in outer space;
- As a world map showing the main languages spoken across the world;
- As a world map showing the spread of the Coronavirus Pandemic across the world;
- As a world map showing political conservatism and liberalism in countries across the world; and
- As a video showing global weather patterns and air jet streams.

Question 1: What do your three ways of viewing the whole of the Planet Earth teach you about your sacred relationship with the whole of life?

Question 2: In what ways is your relationship to Planet Earth sacred? In what ways isn't your relationship sacred?

Question 3: Identify a couple ways you can see the whole of the world in more sacred ways?

Endangered Wildlife Species Exercises

As a first step, identify one endangered wildlife species from the World Wildlife Foundation's (WWF's) website: https://www.worldwildlife.org/species/directory?direction=desc&sort=extinction_status. Read the section about the species on the WWF's website. In the second step, answer these two questions:

Question 1: In what ways do you consider your selected wildlife species to be or not to be sacred? Bear in mind the main points emphasized in *In Sacred Relationship*. If you wish to invest the time, please review the U.S. Fish and Wildlife booklet on endangered species, which you can download in PDF at this website link: https://www.fws.gov/nativeamerican/pdf/why-save-endangered-species.pdf. After you read the booklet, answer Question 2 below:

In Sacred Relationship

Question 2: How can the formation of sacred relationships with endangered wildlife help to protect these species? Write down some compelling thoughts about this question.

Sacred Land or Sacred Site Exercises

The book explores how sacred land is related to the Lakota spiritual worldview. The Badlands, Black Hills, and Devils Tower were examined as Lakota sacred lands. As discussed much earlier in the book, there are three basic views of how land or space becomes known as sacred. The first is through the *identification with special spiritual powers or energies inherent in certain places or lands*, which cause them to be seen as sacred in nature. The second approach is by *"sacralization"* through cultural ritual. This reminds us of the importance of the reenactment of creation stories through spiritual ceremonies. The third way combines the *"inherent spiritual power present within the place"* approach and the *"sacralization through ritual and ceremony"* approach. In the third approach, both ceremony and inherent spiritual presence are seen as necessary for the sacred to exist and be experienced.[280]

What are some specific places you consider to be sacred? They could be:

- Your church or place of worship
- Your home or your parents' or grandparents' home
- Your favorite place in nature
- Your flower garden
- A cemetery or burial place of your ancestors
- A famous religious or spiritual site or shrine
- A place of healing or renewal

Now, it is your turn to identify one specific place that you consider to be personally sacred. Please follow these four steps in completing this exercise:

1. Identify a specific place that you have personally experienced, and that you consider to be sacred land or a sacred site. Briefly describe this place and its location.

---

[280] Chidester, David, and Edward Tabor Linenthal, eds. American sacred space. Indiana University Press, 1995, p. 5-7.

In Sacred Relationship

2. Describe how and why this place is sacred to you. For example, is it sacred because the place has significant religious or spiritual significance, or did you have a special experience in this place?

3. Do you have a sacred relationship with this place? If so, describe the sacredness of the place. If not, why not?

4. If you decided to strengthen your sacred relationship with this place, identify how you might accomplish this.

CHAPTER 8: THREE ILLUSTRATIONS USING THE GROUNDED SPIRITUALITY AND SACRED RELATIONSHIP COMPASS

"Knowledge was inherent in all things. The world was a library."
Luther Standing Bear

A. Introduction

Three applications of the Grounded Spirituality and Sacred Relationship Compass (GSSRC) are discussed in this chapter. Each illustrates the compass's value in coping with *spiritual realities* through sacred relationships. The first application is in the field of *cancer care*, which is a well-known disease that has become the second leading cause of death in the United States and worldwide. Because of its human toll, cancer causes major mental, emotional, and spiritual distress in cancer patients and their families. *Viewed through the lens of sacred relationships, cancer is just as much a spiritual disease of the mind and spirit as it is a physical disease of the body.*

In the second application, we examine how the GSSRC can be used to address spiritual realities related to a newly identified disease, the *Coronavirus Pandemic*, which is a current serious public health threat globally. This new virus is a source of considerable mental, emotional, and spiritual distress. The coronavirus, like cancer, is as much a spiritual disease as a physical disease when we view it from a sacred relationship perspective.

Finally, the third situation is less of a test of the spiritual compass, rather it is an sacred relationship advisory for leaders in business, medicine, education, and government on how the GSSRC can them serve their customers, clients, patients, students, and citizens through sacred relationships. Imagine the added power of patient-centered medicine when we see the patient-caregiver relationship as a sacred relationship. Similarly, the student-teacher relationship shifts to a higher plane with the recognition of the sacred nature of learning and education. The GSSRC transforms the *"service relationship"* in businesses and government agencies when it embodies spiritual realities.

It is important to note that the relevance of the GSSRC's four realms will vary by the issue examined within the compass. Please take this into account in reading the three applications presented in this chapter. Each example is a preliminary test of the GSSRC's value. In each case, only one of many perspectives is tested in the model approach. The primary concern in each case is to show how problems and solutions change when viewed in a sacred relationship context.

As I said earlier in the book, the GSSR Compass is not intended as a rigid framework that must be used in just one way. Instead, the model is a flexible approach to relating to the world in a more sacred way, and as a way to discover and honor the spiritual dimension found in all aspects of life. Individuals, groups, organizations, and communities can use the compass. The compass has value in seeing the *spiritual realities* found in any situation, and it helps us relate to those spiritual realities. By relating in a sacred way, we bring spiritual grounding to our experiences and actions. In our quest to unify the mind, body, and spirit in medicine, complementary medicine, wellness, spirituality, business, education, and many other fields, we have overlooked the obvious source of unification, which is the reality that the whole of life is sacred, and the most important thing we can do is relate to ourselves and everyone else with the recognition of life's sacredness.

B. GSSR Compass Use in Coping with the Spiritual Realities Related to Cancer Care

Cancer is the second leading cause of death in America and worldwide. It was responsible for an estimated 9.6 million deaths worldwide in 2018. Globally, about 1 in 6 deaths is due to cancer.[281] In 2019, an estimated 606,880 people died of cancer in the United States.[282] According to estimates from the Cancer Treatment and Survivorship Statistics, there were more than 16.9 million Americans with a history of

---

[281] Cancer, World Health Organization: https://www.who.int/news-room/fact-sheets/detail/cancer. Accessed February 19, 2020.
[282] National Cancer Institute, SEER Program, How Many People Die of Cancer Each Year?: https://seer.cancer.gov/statfacts/html/common.html, Accessed February 20, 2020.

cancer on January 1, 2019, a number that is projected to reach more than 22.1 million by 2030, based on the growth and aging of the population.[283]

My wife Mary and I work as a team supporting the healing journeys of cancer patients at the Cleveland Clinic Cancer Center in Cleveland, Ohio. Our work lies at the intersection of science and spirituality, which are two worlds that need each other. As complementary medicine and spiritual caregivers to cancer patients, we work with patients from all over the world, who hold diverse religious and spiritual beliefs. Table 2 below contains non-confidential summary data from our 2018-2019 patient care dashboard at the Cancer Center.

Table 2: Complementary Medicine and Spiritual Care Scoreboard Data, Cleveland Clinic Taussig Cancer Center, 2018-2019

Complementary Medicine and Spiritual Care Services[284]	2018	2019	Change
Total Patient Care Encounters	1,363	2,117	754
Reiki	944	1,051	107
Meditation	39	367	328
Meaningful Encounters (Life Purpose Coaching and Spiritual Counseling)	346	667	321
Journaling (Number of patients served, not number of service encounters)	NA	32	32
Patient-Rated Service Satisfaction Score (5 is the best possible score.)	4.91	4.93	0.02

We estimate that about sixty-five (65) percent of our patients use at least one of the services identified in Table 2 above. The data in Table 2 presents a strong case for the need for complementary medicine therapy and spiritual care support for cancer patients.

---

[283] Kimberly D. Miller, Leticia Nogueira, Angela B. Mariotto, Julia H. Rowland, K. Robin Yabroff, Catherine M. Alfano, Ahmedin Jemal, Joan L. Kramer, Rebecca L. Siegel. Cancer treatment and survivorship statistics, 2019. CA: A Cancer Journal for Clinicians, 2019; DOI: 10.3322/caac.21565

[284] **Definitions**: Total patient care encounters is the total of all services identified in Table 2. Reiki is a gentle touch spiritual healing and relaxation therapy. Meditation refers to guided meditation and mindfulness sessions. Meaningful encounters consist of life purpose coaching and spiritual counseling sessions with patients and their families. Journaling is a therapy involving writing about and learning from the cancer journey. Patient satisfaction is an immediate post-service delivery rating provided by patients. Care encounters average 30 minutes in length.

Spiritual Realities for Cancer Patients

The GSSR Compass can help cancer patients and their families cope with spiritual realities related to cancer. What are these spiritual realities? Eight spiritual care realities standout in importance based upon my work with patients:

1. **Each Cancer Patient is a Unique Person and Spiritual Being**: Research indicates that patients wish to be treated as individuals by their healthcare providers.[285] [286] Cancer patients I work with may share a cancer type with many other people, such as breast or lung cancer, but they want to be viewed as individuals and treated as such. There is a strong recognition of and response to this need at Cleveland Clinic Cancer Center.

   While there is growing attention nationally to personalized or precision medicine in cancer care, these efforts do not give adequate recognition to the fact that patient uniqueness has a spiritual dimension in addition to its physical dimension. A recent study published in *Translational Behavioral Medicine* urges greater attention to religion and spirituality in precision medicine.[287] Current research studies on spiritual distress in cancer tend to be too broadly focused on commonly observed issues such as suffering and coping with suffering.[288] We need to understand in more specific terns what role religion and spirituality play in the cancer healing journey. People are unique because of their "spiritual endowments," which goes far beyond what current spiritual distress inventories assess. A 2013 systematic literature review examined 25 widely used spiritual assessment tools. An important conclusion drawn by the researchers is that spiritual assessments are most effective when they are individualized and

---

[285] National Clinical Guideline Centre (UK). Patient Experience in Adult NHS Services: Improving the Experience of Care for People Using Adult NHS Services: Patient Experience in Generic Terms. London: Royal College of Physicians (UK); 2012 Feb. (NICE Clinical Guidelines, No. 138.) 6, Knowing the patient as an individual. Available from: https://www.ncbi.nlm.nih.gov/books/NBK115223/

[286] Entwistle, Vikki A, and Ian S Watt. "Treating patients as persons: a capabilities approach to support delivery of person-centered care." The American journal of bioethics : AJOB vol. 13,8 (2013): 29-39. doi:10.1080/15265161.2013.802060

[287] Yeary, Karen HK, et al. "Considering religion and spirituality in precision medicine." Translational behavioral medicine 10.1 (2020): 195-203.

[288] Martins, Helga, and Sílvia Caldeira. "Spiritual distress in cancer patients: a synthesis of qualitative studies." Religions 9.10 (2018): 285.

explore patent spiritual realities in much greater depth.[289] Spiritual counseling with cancer patients could benefit in the future by working with patients to form sacred relationships with themselves and the world.

2. **Cancer Causes Significant Emotional and Spiritual Distress**: Cancer is a source of significant distress to most patients and their families, who often find difficulty in distinguishing spiritual distress from emotional and mental distress. Patients often experience emotional, mental, and spiritual distress as interrelated, and often difficult to separate. This is the reason why we treat the *"whole patient"* in our work.

3. **The Importance of Religion and Spirituality Varies Among Patients**: Religion and spirituality are important to most, but not all, cancer patients. For some patients, their cancer experience has made religion and spirituality much more important in their lives, especially in confronting challenging existential questions about life and death. For a few patients, their cancer experience has caused them to question the value of their religion or spirituality.

4. **Religion and Spirituality Mean Different Things to Patients**: Spirituality and religion often mean different things to different patients. When patients talk about their religion, they usually refer to their religious affiliation and the church they attend. On the other hand, spirituality often refers to the patient's personal relationship with God or the Divine. Most patients see religion and spirituality as compatible with each another.

5. **Spiritual Beliefs are Often Private and Always Personal**: Many patients are very private about their religious and spiritual beliefs, because they do not want to feel pressured to defend their beliefs. I find that spiritual conversations with patients most often evolve out of how they see cancer affecting their lives, families, and work. Spiritual issues are not usually where patient conversations begin

---

[289] Lucchetti, Giancarlo et al. "Taking spiritual history in clinical practice: a systematic review of instruments." Explore 9 3 (2013): 159-70.

because they are complex and personal. Patients are willing to explore how their religion or spirituality can help them more directly during their cancer healing journeys. This is a good sign for grounded spirituality.

6. **Medical Care and Spiritual Care are Sometimes Seen as Separate**: While most patients see the importance of their religion and spiritualty in their cancer healing journey, many also see medicine and religion/spirituality as different realms of their lives. In short, they look to Cleveland Clinic for their cancer medical care, and they look primarily to their church for their spiritual care. Sometimes patients have trouble connecting these two realms.

7. **Complementary Medicine Therapies are Viewed as Beneficial Regardless of Religious and Spiritual Affiliation:** The majority of our patients view Reiki, meditation, mindfulness, yoga, and other complementary medicine services as beneficial, and most do not see these therapies as inconsistent with their religious and spiritual beliefs. Patients who are very private about their cancer journey, and those who are most religiously conservative, are least likely to use complementary medicine and spiritual care services.

8. **Frequently Asked Spiritual Questions by Patients**: Patients with whom I have an ongoing relationship are much more likely to ask personal spiritual questions. First-time patients and those I see infrequently are less likely to ask spiritual questions. Here are some of the most common spiritual questions patients ask. Because my communication with patients about these questions depends upon the patient's spiritual needs, I will avoid generic answers to these questions here. The top spiritual questions I receive are these:

   - **Concerns for Loved Ones and Close Friends**: Patients are very concerned about the welfare of their family members and close friends in light of their cancer. Their questions related to this concern are: 1) What can I do to help my loved ones experience less emotional, mental, and spiritual pain and suffering during my cancer journey and afterwards if I die; and

2) What can I do to help myself feel less pain and suffering about leaving my loved ones and friends?

- **Faith Healing**: Is there any scientific evidence of spiritual healing of cancer through prayer and faith healing? While most patients asking this question are referring to the physical healing of cancer through religious and spiritual belief and practice, some also are inquiring about emotional, mental, and spiritual healing.

- **Spiritual Distress and Spiritual Growth**: Can I relate in a better spiritual way to my cancer? In other words, can I spiritually cope with my sense of mortality brought on by my cancer, and also how can I grow spiritually from my cancer experience?

- **Spiritual Causes of Cancer**: Are there spiritual reasons for my cancer? Another version of this question is: Does God punish people with cancer? Phrased another way, why do I have cancer?

- **Spiritualty's Role in Cancer Outcomes**: This is a different question than the faith healing question above. Is there any evidence that religion and spirituality can improve cancer outcomes? Is this evidence real, or is it explained by the placebo effect?[290] If I put more faith in my religion and spirituality, will that improve my cancer outcome?

- **Life Meaning and Uncertainty**: Cancer challenges many patients' sense of life meaning, especially where cancer causes major lifestyle changes. Cancer is a major source of

---

[290] In medicine, a placebo is anything that seems to be a "real" medical treatment, but is not. Placebos could include pills, shots, or some other type of "fake" treatment. All placebos share in common that they do not contain an active substance meant to affect health. Placebos are found to help health outcomes in some instances, but not all. In spirituality and religion, a placebo can be beliefs in faith healing, the power of prayer, and the effects of meditation, yoga, Reiki, and other complementary therapies with a basis in spiritual belief. The placebo effect is commonly referred to as *"mind over body or matter."*

uncertainty in the lives of cancer patients and their families. Related questions are: 1) How can my religion and spirituality help me restore my sense of life meaning and purpose; and 2) How can my religion or spirituality help me manage cancer uncertainties?

- **Dying, Death and the Afterlife**: For many seriously ill cancer patients, their questions about dying, death, and the spiritual afterlife become more important. Common patient questions are: 1) How can I die a "*good death*" when the time comes; that is how can I die with less pain and more dignity and meaning? 2) Since nobody, while they are still alive, knows what happens in the spiritual afterlife, do my ideas about the spiritual afterlife make any sense, and do my ideas reflect new understandings of the afterlife? and 3) Does it really matter what I believe about death and the afterlife since nobody really knows what happens?

Each of these questions has complexities and nuances, which are best understood in a specific patient context. Active listening and helping the patient find their own answers are the two most effective spiritual counseling strategies in my experience. However, in some cases patients are looking for ideas beyond their own. Some appreciate receiving a list of options they can review and consider. Spiritual caregivers must be highly sensitive to avoid adding to a patient's distress.

GSSRC Applied to the Patient Perspective in Cancer Caregiving

In this section, we explore how the GSSR Compass can be applied to cancer caregiving. *The relationship perspective given primary attention in this application is the patient in the patient-caregiver relationship.* This means that the test of the application is to examine sacred relationship formation by the patient. Caregiver refers to: 1) personal caregivers such as family members and friends; and 2) clinical caregivers such as physicians, nurses, radiology therapists, psychosocial caregivers, physical and occupational therapists, complementary medicine caregivers, and spiritual caregivers involved in cancer medicine.

For starters, let us look at how each GSSRC realm is relevant in the cancer care example. The four realms are <u>redefined</u> in the following way in the cancer care example:

1. Patient's inner reality (mind, spirit, soul);
2. Patient's body (both physical and energy);
3. Other people (personal and clinical/spiritual caregivers); and
4. Patient's whole life (family, friends, work colleagues, others).

Figure 15: Cancer Patient Sacred Relationships

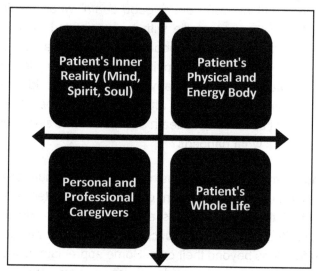

Sacred relationships are possible and beneficial in all aspects of the cancer patient's healing journey.

Patient's Inner Reality

Volumes have been written in many disciplines over the years about human inner reality, its nature, and functioning. My working definition of human inner reality is this: We experience ourselves as having an inner reality, or an inner experienced reality, which we term the inner self. We associate our thoughts, feelings, sensations, beliefs, and other inner experiences with the workings of our inner self. We experience our "inner reality" as separate from what we term our "outer reality." The inner self is a subjective reality that is phenomenological in nature. These experiences lead us to conclude that an inner self exists with our inner reality. Is the inner self fixed in identity or variable? This is another highly debated issue. Within Buddhist philosophy, the self is viewed as everchanging, and not fixed. Some neuroscientists concur with this "changing self" notion.[291] I do not see our inner lives as fixed in nature. Instead, they are flowing and changing as "processes" unfolding within us.

The inner self exists within our "experience" of the inner dimension of our life, which is where our mental, emotional, and spiritual processes take place. To know your inner self is to know your purpose, values, vision, goals, motivations, and beliefs. Knowing our true inner selves requires a high level of self-reflection, contemplation, and introspection.

Three interrelated elements or aspects comprise the inner self. These are the:

- **Soul**: This is our deepest spiritual bedrock, which is our spiritual whole. This is where our eternal natural divinity resides. The soul is in closest proximity to God. Mind and spirit flow from the soul.
- **Spirit**: This is best understood as our "personhood," which some call the spiritual personality.
- **Mind:** This is our mental and emotional consciousness, which is our everchanging awareness of ourselves, others, and the rest of the world.

Patient's Body

Cancer is a terrifying disease that attacks the body on many fronts, which is the main concern in scientific medicine, but cancer also attacks the mind and spirit. While our scientific understanding of cancer has grown substantially over the years, there remains a great deal to learn about the science of cancer in the future. *In Sacred Relationship* respects the scientific medicine perspective of cancer, while advocating for greater attention inside and outside medicine in dealing with the spiritual realities associated with cancer. These spiritual realities were discussed earlier.

Let us start this section by looking at the National Cancer Institute's definition of cancer: "*A term for diseases in which abnormal cells divide without control and can invade nearby tissues. Cancer cells can also spread to other parts of the body through the blood and lymph*

---

[291] Dahl, Cortland J., Antoine Lutz, and Richard J. Davidson. "Reconstructing and deconstructing the self: cognitive mechanisms in meditation practice." Trends in cognitive sciences 19.9 (2015): 515-523.

systems. There are several main types of cancer. Carcinoma is a cancer that begins in the skin or in tissues that line or cover internal organs. Sarcoma is a cancer that begins in bone, cartilage, fat, muscle, blood vessels, or other connective or supportive tissue. Leukemia is a cancer that starts in blood-forming tissue, such as the bone marrow, and causes large numbers of abnormal blood cells to be produced and enter the blood. Lymphoma and multiple myeloma are cancers that begin in the cells of the immune system. Central nervous system cancers are cancers that begin in the tissues of the brain and spinal cord. Also called malignancy."

As we contemplate sacred relationships with the body in coping with cancer, these ideas may be helpful guides:

1. The physical body is an integrated functioning whole, which means we must be in sacred relationship with our whole body and not just the part(s) with cancer.

2. The body is never separate from the mind and spirit, which play roles in the presence and growth of cancer, as well as our recovery, cure, and healing. We must be in sacred relationship with our minds and spirits, as well as our bodies.

3. All diseases, including cancer, have intelligence, which means we must relate to the intelligence found in our abnormal cell growth. We must be in sacred relationship to our cancer and its intelligence.

Patient's Personal and Professional Caregivers

I made this point earlier: Our relationships define us. In that regard, how we relate to family, friends, work colleagues, and others influences our healing. Others' love and support are vital to our healing, and therefore we must be in sacred relationship with others to receive their love and support in a way that promotes our resilience and healing. The same is true in our relationships with our professional caregivers, who are the best sources of the knowledge and treatments that we need to heal. Gratitude is a cornerstone to healing. Being in "grateful relationship" to others is central to our healing on all levels.

Patient's Whole Life

We live within the complex and everchanging web of life, and therefore we must be in sacred relationship with the web of life. We must honor the web of life and recognize its importance to cancer healing. How does the web of life affect us? For one, cancer has genetic roots, which carries across our family history. Environmental pollution is a source of cancer, including our direct and indirect exposure to environmental toxins. Lifestyle has a great deal to do with the occurrence of cancer. Our lifestyles are rooted in our families, immediate neighborhood, and larger social, economic, political, and cultural contexts. For these reasons, cancer recovery, cure, and healing depend upon our relationship to the whole of life.

Understanding and Fostering Sacred Relationships in Cancer Care

Sacred relationship-building can help cancer patients with each of the four realms identified in Figure 15 earlier. In an overall sense, grounded spirituality can help patients use their cancer journey to cultivate *"sacred relationships"* with themselves, others, and their cancer healing process. Cancer can be a spiritual teacher, which is the topic of my 2017 book, containing poems and narratives about the spiritual realities of cancer.[292] My meaningful encounters with patients indicate that their cancer experiences have taught many of them important spiritual lessons.

These four realms are the:
1. Patient's inner reality;
2. Patient's physical and energy bodies;
3. Patient's personal and professional caregivers; and
4. Patient's whole life.

Each of these four realms is examined in the context of the four phases of the cancer journey in Table 3 below. Life before diagnosis refers to the patient's life before they know they have cancer. Concern about cancer is usually very low in phase 1. Concern about cancer is very high in phases 2 and 3. Concern about cancer is low in phase 4, if treatment is effective, and it remains high if treatment is not effective. The GSSRC helps us establish sacred relationship is each of the four realms.

---

[292] Iannone, Donald. T., Cancer as Spiritual Teacher: Poems on Walking the Healing Path, Kindle Direct Publishing, August 2017.

In Sacred Relationship

Table 3: Understanding Sacred Relationships During the Stages of Cancer Journey

Cancer Journey Phase	Patient's Inner Reality	Patient's Physical and Energy Body	Patient's Personal & Professional Caregivers	Patient's Whole Life
1. Life before diagnosis: Time free of cancer and, also conditions giving rise to cancer.	Patient's relationship to their inner life before cancer, and possible roles of inner realities giving rise to cancer.	Patient's relationship to their body before cancer, and possible roles of body realities giving rise to cancer.	Patient's relationship to other people before cancer, and possible roles of other people in the life before cancer.	Patient's relationship to the whole of life before cancer, and possible roles of their whole of life in cancer.
2. Diagnosis: Type and stage of cancer and prescribed treatment.	Patient's relationship to their inner life during diagnosis.	Patient's relationship to their body during diagnosis.	Patient's relationship to other people during diagnosis.	Patient's relationship to their whole of life during diagnosis.
3. Treatment: Type of treatment prescribed and its success in combatting cancer.	Patient's relationship to their inner life during treatment.	Patient's relationship to their body during treatment	Patient's relationship to other people during treatment.	Patient's relationship to the whole of life during treatment.
4. Life after treatment: Survivorship and future quality of life.	Patient's relationship to their inner life after treatment.	Patient's relationship to their body after treatment.	Patient's relationship to other people after treatment.	Patient's relationship to the whole of life after treatment.

Each of the four phases is an opportunity to cultivate sacred relationships. Table 4 below identifies sacred beliefs and practices that can be beneficial in any of the four phases.

One important, often overlooked point in how we think about cancer and other serious diseases is that we must *respect* diseases because: 1) they are an intimate part of us, and how we treat them affects how we treat our personal wholeness; and 2) cancer reminds us of the consequences of our relationships with life. They remind us that everything in our lives is a result of something, including how we view life, how we live, and whether learn from our life experiences and traumas.

Table 4: Spiritual Practices for Cultivating Sacred Relationships During the Cancer Journey

Cancer Journey Phase	Patient's Inner Reality	Patient's Physical and Energy Body	Patient's Personal & Professional Caregivers	Patient's Whole of Life
1. Life before diagnosis: Conditions giving rise to cancer.	Gratitude for good health and blessings for balanced mind, body, and spirit.	Gratitude for good health and blessings for balanced mind, body, and spirit.	Gratitude for and blessings upon family, friends, and one's extended web of relations.	Gratitude for and blessings upon the whole of life that sustains us all.
2. Diagnosis: Type and stage of cancer and treatment plan.	Inner gratitude and strength in dealing with uncertainty and trauma.	Bodily gratitude and strength in dealing with uncertainty and trauma.	Gratitude and strength personal and professional caregivers.	Gratitude and strength for the whole of life sustaining us.
3. Treatment: Type of treatment and its success	Inner gratitude and strength through the treatment process.	Bodily gratitude and strength through the treatment process.	Gratitude and strength for caregivers through the treatment process.	Gratitude and strength for the whole of life through the treatment process.
4. Life after treatment: Survivorship and future quality of life.	Inner gratitude and strength for healing and wholeness during survivorship.	Bodily gratitude and strength for healing and wholeness during survivorship.	Gratitude and strength for caregivers for wholeness and healing during survivorship.	Gratitude and strength for the whole of life through survivorship.

The GSSRC helps the cancer patient and their caregivers to relate to the sacred realities of all phases of the cancer journey, including the phases before and following cancer diagnosis and treatment. The keys are: 1) honoring the precious gift of life; 2) keeping one's life in sacred relationship throughout the journey; and 3) fostering gratitude and strength throughout the journey.

C. GSSR Compass Use in Coping with Spiritual Realities Related to the Coronavirus Pandemic

Introduction

The Coronavirus Pandemic is a current source of great mental, emotional, and spiritual distress to people worldwide. The Grounded Spirituality and Sacred Relationship Compass (GSSRC) can help us cope with the spiritual realities surrounding this major public health threat. This section provides a test application of the GSSRC as a spiritual strategy in dealing with this major public health crisis.

About the Coronavirus (COVID-19)

What is the coronavirus disease 2019? According to the U.S. Center for Disease Control (CDC), the Coronavirus disease 2019 (COVID-19) is a respiratory illness that can spread from person to person. The virus that causes COVID-19 is a novel coronavirus that was first identified during n investigation into an outbreak in Wuhan, China.[293]

As of March 14, 2020, there were 143,000 confirmed worldwide cases of people infected by the virus; of which 82,000 were in China and about 1,700 in the U.S. The CDC relies on several external forecast sources on the future spread of the disease. One of those outside groups, the Laboratory for the Modeling of Biological and Socio-technical Systems at Northeastern University prepared future spread scenarios based upon high, medium, and low surveillance and intervention. The Northeastern model estimates that by April 30, 2020:[294]

- Under the high surveillance and intervention scenario, the disease could infect 160,000 Americans.
- Under the medium surveillance and intervention scenario, infections could grow to 356,000.
- Finally, under the low surveillance and intervention scenario, 1.4 million Americans could be infected.

Even more frightening is the CDC's "worst-case" forecast of U.S.

---

[293] CDC, About the Coronavirus: https://www.cdc.gov/coronavirus/2019-nCoV/index.html. Accessed March 15, 2020.
[294] Wilson, Chris, Exclusive: Here's How Fast the Coronavirus Could Infect Over 1 Million Americans, Time Magazine, March 12, 2020: https://time.com/5801726/coronavirus-models-forecast/. Accessed March 15, 2020.

COVID-19 infections and deaths. An unpublished CDC forecast, which was viewed by the New York Times suggests that: *"Between 160 million and 214 million people in the United States could be infected over the course of the epidemic, according to a projection that encompasses the range of the four scenarios. That could last months or even over a year, with infections concentrated in shorter periods, staggered across time in different communities, experts said. As many as 200,000 to 1.7 million people could die."*[295]

Spiritual Realities Surrounding the Coronavirus Pandemic

Crisis throughout history has brought out the best and worst in people worldwide. My deepest wishes related to the Coronavirus Pandemic are that: 1) we stem its spread quickly; 2) we find improved treatment and cure for the virus; and 3) we learn important spiritual lessons from the crisis about interdependence, personal and social responsibility, compassion, and working together for the good of all.

The *In Sacred Relationship* spiritual compass can help us to navigate this crisis individually and collectively. The graphic in Figure 16 below shows how the GSSRC applies to the Coronavirus Pandemic.

Figure 16: GSSR Compass Applications to Coronavirus Pandemic

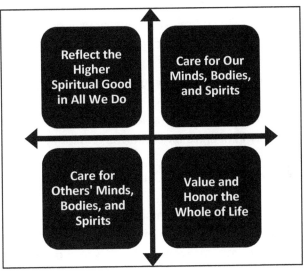

---

[295] Fink, Sheri, Worst-Case Estimates for U.S. Coronavirus Deaths, New York Times, March 14, 2020: https://www.nytimes.com/2020/03/13/us/coronavirus-deaths-estimate.html. Accessed March 15, 2020.

Alignment with Higher Spiritual Purpose

Sacred Relationship Objective
Reflect the Higher Spiritual Good in all that we do.

Sacred Alignment Steps
Each of us is encouraged to align our attitudes and actions related to the Coronavirus Pandemic with our own understanding of the Higher Spiritual Good. We achieve sacred relationship when we:

1. Honor life's sacredness found in everyone living everywhere;
2. Use the crisis to strengthen the power of faith and foster spiritual growth in ourselves and others; and
3. Act in line with spiritual goodness and scientific understanding.

Role of Faith Communities and the Faithful
Faith communities across the world must do their part in helping their congregations and members through this crisis.[296] Faith communities, like all other sectors of society, must observe the laws, rules, and advice provided by health and governmental officials charged with containing this pandemic. That includes voluntarily closing churches and other places of worship, where deemed necessary.

Moreover, faith communities must work together through interfaith dialogue and collaboration to address larger surrounding and shared spiritual realities related to the Coronavirus Pandemic. This is a time for building upon our spiritual common ground that brings all of us together to act with knowledge, compassion, reason, and kindness. We must be guided by authentic and demonstrable spiritual purpose in making the right decisions in this time of crisis for ourselves and for the whole of the world.

Personal Responsibility for Ourselves

Each of us must accept personal responsibility for ourselves in the context of our families, local communities, nations, and global community. When we act out of fear and anger, the potential exists for us to see the

---

[296] CDC, Coronavirus Disease (COVID 2019), Resources for Community and Faith-based Leaders: https://www.cdc.gov/coronavirus/2019-ncov/community/organizations/index.html. Accessed March 15, 2020.

world in zero-sum terms, leading us to take care of our needs at the expense of others. Hoarding behavior is an example of zero-sum action. Similarly, we must do our best at "sitting things out" and taking self-imposed quarantine more seriously to reduce the spread of the virus.[297] It is irresponsible of us to place ourselves and others at risk by not following CDC guidelines about public health protection. And if we are infected by the virus, we must do everything in our power to not expose others.[298]

Social Responsibility for Others

We have a social responsibility to others in how we act relative to the Coronavirus Pandemic. Our social responsibility is four-fold:

1. Recognize our own role in dealing with this crisis
2. Show compassion and empathy toward others
3. Treat others with kindness and respect
4. Accept personal responsibility and avoid blame

The GSSR spiritual compass encourages us to hold the following groups of people in sacred relationship in the context of this pandemic:

1. Our families
2. Our friends and neighbors
3. Our work colleagues
4. Our patients, customers, and clients
5. Our healthcare providers
6. Our government workers (all levels)
7. Our many personal service providers, such as our landscapers, housekeepers, grocery store clerks, delivery service drivers, financial advisors, pet sitters, others who make our lives easier.
8. Our spiritual caregivers (all types)
9. Our elected officials (all levels)

---

[297] Nunneley, Chloë E., et al., Slowing the spread of coronavirus: Why some people aren't social distancing, ABC News, March 13, 2020: https://abcnews.go.com/Health/slowing-spread-coronavirus-people-social-distancing/story?id=69574388. Accessed March 15, 2020.

[298] Bumbaca, Chris, Rudy Gobert's coronavirus message: 'Wish I would have taken this thing more seriously, USA Today, March 15, 2020: https://www.usatoday.com/story/sports/nba/jazz/2020/03/15/utah-jazzs-rudy-gobert-wishes-he-took-coronavarius-more-seriously/5054253002/. Accessed March 15, 2020.

In Sacred Relationship

10. The news media that keeps us informed
11. Our business and community leaders
12. Anyone else who adds value to our lives.

Responsibility for the Whole of Life

Life is a precious gift in all its forms. We must do our best to honor and value life in how we relate to the coronavirus. This requires each of us to relate to the whole of life in a sacred way. To ensure that we honor the whole of life, it is helpful to reflect upon this whole at the various levels at which we encounter it. This includes:

1. Personal family household
2. Extended family
3. Friends network
4. Place of employment
5. Neighborhood and local community
6. Larger surrounding areas: region; state; nation; and worldwide.

Concluding Thoughts on Sacred Relationships with the Coronavirus Pandemic Situation

The GSSR spiritual compass reminds us to align ourselves in sacred relationship to COVID-19 and everyone touched by the virus. The compass reminds us to reflect the Higher Spiritual Good in all we do, and it reminds us see the "big life picture" beyond ourselves. Our relationships define us. When we relate to life in a sacred way, especially during times of crisis, we grow richer in a spiritual and human sense.

While this application of the spiritual compass has been cursory, it is sufficient to show us the compass's power in helping us do what is good and right at this time.

D. Advice on Using the GSSR Spiritual Compass in Serving Patients, Customers, Clients, and Citizens

Our service to others becomes more valuable to those we serve and ourselves when it is approached in sacred relationship. In this way, we honor the life behind the customer, client, citizen, congregant, and

In Sacred Relationship

patient we serve. This deepens our service relationship beyond the usual rules of engagement and service. The GSSR spiritual compass prepares us to see the spiritual purpose in whatever work we perform. Figure 17 below calls attention to how the compass applies to our service to others.

Figure 17: GSSR Compass Application to Our Service to Others

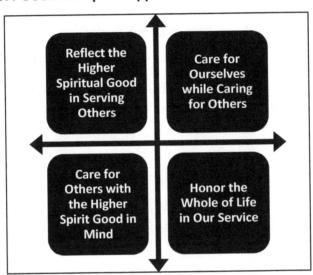

Reflect the Higher Spiritual Good in Serving Others

This means serving others with the Higher Spiritual Good in mind. Our service in this context takes on two forms of value. The first is the *literal value*, which is the visible business, medical, economic, financial, and utility value we provide to others. The second is the *symbolic* value, which is the spiritual value we honor in others and bring to others through our service. The latter is the new value-added we provide. The symbolic value is the recognition of the spirit and soul underlying our patients, customers, clients, congregants, and any others we serve. This is possible when we serve others in sacred relationship.

Caring for Ourselves while Caring for Others

It is vital that we care for ourselves as we serve others. Our lives are also sacred, and therefore must be honored. The bottom line is that we can care better for others when we take care of ourselves.

Caring for Others with the Higher Spiritual Good in Mind

This is pretty much covered under "Reflecting the Spiritual Good in Serving Others." The one additional idea to add here is that everyone becomes a "spiritual caregiver" when we approach our work in sacred relationship.

Honoring the Whole of Life in Our Service

Beyond our responsibilities to those we serve, our work should have the best interests of the whole of life in mind. Our work should sustain the whole of humanity and Planet Earth. Each of us belongs to the whole.

CHAPTER 9: CONCLUDING THOUGHTS

"You are all my relations, my relatives, without whom I would not live. We are in the circle of life together, co-existing, co-dependent, co-creating our destiny. One, not more important than the other. One nation evolving from the other and yet each dependent upon the one above and the one below. All of us a part of the Great Mystery."
Mitakuye Oyasin – Lakota Prayer

I began writing this book in advance of the Coronavirus Pandemic. From the start, I felt a deep sense of urgency about the book, believing that it contained a message that needed to be heard *now*. I understand now why it was important that I complete this book quickly and make it available to others.

Despite the pain and suffering it has caused, the Coronavirus Pandemic is a valuable and timely reminder of the sacredness and interconnectedness of life everywhere across the world. *In Sacred Relationship* shows us how we can respond effectively and compassionately to the many complex realities surrounding this crisis. The book also shows us that it matters <u>how</u> we respond to developing events in our lives, and it encourages all of us to work together to contain the spread of the virus, while bearing in mind that all of life is sacred.

In Sacred Relationship was inspired by the concept of *"all my relations"* in Lakota spirituality. The Lakota prayer found above helps us understand the importance of *"all my relations"* as a way of understanding and responding to the major challenges facing today's world. *"All my relations"* is a spiritual idea that the world needs today.

In review, the first third of the book provides an organization to the book. Part 2 imparts an understanding of Lakota spirituality, including the role of sacred land within the Lakota spiritual worldview. While some readers may be inclined to jump to the book's second half, I encourage everyone to read and contemplate what it has to say. The last third of the

In Sacred Relationship

book provides a new spiritual compass to help people of all faiths and those with no declared faith to improve their sacred alignment with God or the Divine, themselves and others, and the whole of life. This compass is based upon the Lakota idea of *"all my relations."*

As the reader of this book, I encourage you to work with the exercises provided, learn from them, and strengthen your own sacred relationship with the world. Please allow the three applications of the GSSR compass to deepen your understanding of the compass and how it can serve as a tool for aligning with the sacred.

I am eager to receive feedback from you as the reader. Please share with me what you learned. Together, let's grow a stronger circle of compassion and trust across the world.

ABOUT THE AUTHOR

Don Iannone is an author, complementary medicine therapist, and spiritual caregiver. He works at the Cleveland Clinic Cancer Center in Cleveland, Ohio. Don is also the Managing Director of *kosmos*, a private consulting and research company working in healthcare, the spiritual and religious sector, and economic and community development. Currently, Don's chief area of interest is in advancing the role of grounded spirituality in medicine and other aspects of society. For three decades, Don worked in the economic development and environmental quality fields. He led economic development and environmental centers at Cleveland State University for almost 15 years. Don has worked extensively with American Indian tribes and conducted research on Native American spirituality. He transitioned into healthcare during the last decade. He has served as a consultant to leading healthcare organizations, including Cleveland Clinic, Allegheny Health Network, and the Cherokee Nation of Oklahoma. He is the author of over 20 academic and professional articles and 11 books. Don holds a Bachelor's Degree in Anthropology, Master's Degree in Mind-Body Medicine, Master of Divinity Degree, and a Doctor of Divinity Degree. Don is an ordained interfaith minister, and he is certified as a Reiki Master, Meditation Guide, Mindfulness Practitioner, and Life Purpose Coach. Don lives in Cleveland, Ohio with his wife Mary. He has two sons, Jeff, and Jason, and two grandsons, Evan, and Griffin. Don grew up in Eastern Ohio in Martins Ferry and St. Clairsville. Contact me: http://www.donaldiannone.com.

REFERENCES AND SOURCES CITED

A. Books and Articles

Aldred, Lisa. "Plastic shamans and astroturf sun dances: New Age commercialization of Native American spirituality." American Indian Quarterly 24.3 (2000): 329-352.

Arden, Harvey, and Steve Wall. Wisdomkeepers: Meetings with Native American Spiritual Elders. Beyond Words Publishing, Inc., 4443 NE Airport Rd., Hillsboro, OR

Basso, Keith, H., Wisdom Sits in Places, University of New Mexico Press, 1996.

Benton, Rachel, C., et al. "Baseline mapping of fossil bone beds at Badlands National Park." 6th Annual Fossil Resources Conference. 2001.

Black Elk, N. (1980). Wiwanyag Wachipi: The sun dance. In J. E. Brown (Ed.), The sacred pipe: Black Elk's account of the seven rites of the Oglala Sioux (pp. 67-100). Norman: University of Oklahoma Press.

Elk, Wallace Black, and William S. Lyon. Black Elk: The sacred ways of a Lakota. HarperCollins Publishers, 1990.

Bonvillain, Nancy. The Teton Sioux. Infobase Publishing, 2005.

Bowman, Marion. "The noble savage and the global village: Cultural evolution in new age and neo-pagan THOUGHT." Journal of Contemporary Religion 10.2 (1995): 139-149.

Brady, Joel, "Land Is Itself a Sacred, Living Being": Native American Sacred Site Protection on Federal Public Lands Amidst the Shadows of Bear Lodge, 24 American Indian Law Review, 2000, p. 153 -185.

Chidester, David, and Edward Tabor Linenthal, eds. American sacred space. Indiana Univ. Press

Cousins, Emily. "Mountains made alive: Native American relationships with sacred land." CrossCurrents (1996): 497-509.

Churchill, Ward. "The black hills are not for sale: A summary of the Lakota struggle for the 1868 treaty territory." The Journal of Ethnic Studies 18.1 (1990): 127.

Deloria, Vine, Spirit and Reason, Golden, CO: Fulcrum Publishing, 1999, p. 323-337.

Deloria, Vine. God is red: A native view of religion. Fulcrum Publishing, 2003, p. xv.

DeMallie, Raymond J. "The Lakota Ghost Dance: An Ethnohistorical Account." Pacific Historical Review 51.4 (1982): 385-405.

Diotalevi, Robert Nicholas, and Susan Burhoe. "Native American Lands and the Keystone Pipeline Expansion: A Legal Analysis." Indigenous Policy Journal 27.3 (2017).

Elkins, David N. "Psychothe Rapy and Spirituality: Toward a Theory of the Soul." Journal of Humanistic Psychology, vol. 35, no. 2, Apr. 1995, pp. 78–98,

Ellerby, Jonathan H. "Spirituality, holism and healing among the Lakota Sioux, towards an understanding of Indigenous medicine." (2000).

Erdoes, R., and A. Ortiz, eds. American Indian myths and legends. Pantheon, 1984, p. 50-51.

Erdoes, Richard. Crow Dog: Four generations of Sioux traditional healers. New York: HarperCollins, 1995.

Foster, Charles. Wired For God: The biology of spiritual experience. Hachette UK, 2011.

Flood, Reneé S. Lost Bird of Wounded Knee: Spirit of the Lakota. NY: Scribner, 1995.

Garroutte, Eva M., et al. "Religiosity and spiritual engagement in two American Indian populations." Journal for the Scientific Study of Religion 48.3 (2009): 480-500.

Garroutte, Eva Marie, et al. "Religio-Spiritual Participation in Two

American Indian Populations." Journal for the scientific study of religion 53.1 (2014): 17-37.

Gilmartin, M. (2009). "Colonialism/Imperialism". Key concepts in political geography, (pp. 115–123). London: Sage Publishing

Glacken, C., J., Traces on the Rhodian Shore. Berkeley, CA: University of California Press, 1967, p. 35

Glick, Ashley A. "The Wild West Re-Lived: Oil Pipelines Threaten Native American Tribal Lands." Vill. Envtl. LJ 30 (2019): 105.

Goodman, Ronald, Lakota Star Knowledge, Archeoastronomy; College Park, Md. Vol. 12, (Jan 1, 1996): 140.

Hanson, J. R., and S. Chirinos. 1991. "Ethnographic Overview and Assessment of Devils Tower National Monument." University of Texas, Arlington.

Harvey, Graham. Animism: Respecting the living world. Wakefield Press, 2005, p. xi.

Iannone, Donald, T., Cancer as Spiritual Teacher, Poems on Walking the Healing Path, Kindle Direct Publishing, August 2017.

Jewell, Benjamin, "Lakota Struggles for Cultural Survival: History, Health, and Reservation Life" (2006). Nebraska Anthropologist.

Jones, Lindsay, and Mircea Eliade. The encyclopedia of religion. Macmillan Reference USA, 2005, Lakota Religious Traditions, p. 5295-5298.

Joseph R. The limbic system and the soul: evolution and the neuroanatomy of religious experience. Zygon. 2001; 36:105-136.

Kong, L. (1990) Geography and religion; trends and prospects. Progress in Human Geography 14; 355-371

Krishnan, Venkat R. "The impact of transformational leadership on followers' duty orientation and spirituality." Journal of Human Values 14.1 (2008): 11-22.

Kurkiala, Mikael. "Objectifying the past: Lakota responses to Western historiography." Critique of anthropology 22.4 (2002): 445-460.

LaDuke, Winona, In the Time of the Sacred Places in The Wiley Blackwell Companion to Religion and Ecology, First Edition. Edited by John Hart, 2017, John Wiley & Sons, p. 73-74.

LaDuke, Winona. All our relations: Native struggles for land and life. Haymarket Books, 2017.

Lass, William E. (2000). Minnesota: A History. New York, NY: W. W. Norton & Company.

LaPointe, James, and Louis Amiotte. Legends of the Lakota. Indian Historian Press, 1976.

McCullers, Carson. The heart is a lonely hunter. Houghton Mifflin Harcourt, 2010.

McLoughlin, Lisa A. "US Pagans and Indigenous Americans: Land and Identity." Religions 10.3 (2019): 152.

McGaa, Ed., Mother Earth Spirituality: Native American Paths to Healing Ourselves and Our World. 1990.

Mengden IV, Walter H. "Indigenous People, Human Rights, and Consultation: The Dakota Access Pipeline." American Indian Law Review 41.2 (2017): 441-466.

Miller, Tracy. US Religious Landscape Survey Religious Beliefs and Practices: Diverse and Politically Relevant. 2008.

Muller, Rene, J., Neurotheology: Are We Hardwired for God, Psychiatric Times, May 2, 2008, accessed on January 8, 2020 at https://www.psychiatrictimes.com/neurotheology-are-we-hardwired-god

Nabokov, Peter. Where the lightning strikes: The lives of American Indian sacred places. Penguin, 2007, p.5.

Oakes, Jill. "Sacred Lands: Aboriginal World Views." Claims and Conflicts, Edmonton, Alberta: Canadian Circumpolar Institute, University of Alberta (1988).

Osbon, D. K. (1991). Reflections on the Art of Living. A Joseph Campbell Companion. United States of America: Harper Perennial, p. 180.

Ostler, Jeffrey, The Lakotas and the Black Hills, The Struggle for Sacred Ground, Penguin Books, NY, 2010.

Otto, Rudolph. "The idea of the holy. (Harvey, J., Trans.) New York." (1958).

Palmer, Parker J. A hidden wholeness: The journey toward an undivided life. John Wiley & Sons, 2009.

Posthumus, David C. "All My Relatives: Exploring Nineteenth-Century Lakota Ontology and Belief." Ethnohistory 64.3 (2017).

Posthumus, David. "Transmitting Sacred Knowledge: Aspects of Historical and Contemporary Oglala Lakota Belief and Ritual," PhD dissertation, University of Indiana, 2015

Powers, William, K., Oglala Religion, Bison Book Printing, 1982.

Pritzker, Barry M. A Native American Encyclopedia: History, Culture, and Peoples. Oxford: Oxford University Press, 2000, p. 329.

Rickett, Allyn, in Guanzi: Political, Economic, and Philosophical Essays from Early China: A Study and Translation. Volume II. Princeton University Press, 1998, p. 106.

Robbins, Jeffrey W. "In search of a non-dogmatic theology." CrossCurrents (2003): 185-199.

Roberts, Richard L., et al. "The Native American medicine wheel and individual psychology: Common themes." Journal of Individual Psychology 54 (1998): 135-145.

Rood, David, S., and Taylor, Allan, R., 1996. Sketch of Lakhota, a Siouan Language. In Ives Goddard (ed.), Languages, 440-482. Washington, D.C.: Smithsonian Institution.

Sanchez, Tony R. "The depiction of Native Americans in recent (1991–2004) secondary American history textbooks: How far have we come?."

Equity & Excellence in Education 40.4 (2007): 311-320.

Schell, Herbert S. (2004). History of South Dakota: South Dakota State Historical Society Press.

Sharpley, Richard, and Priya Sundaram. "Tourism: A sacred journey? The case of ashram tourism, India." International journal of tourism research 7.3 (2005): 161-171.

Silko, Leslie Marmon. Language and literature from a Pueblo Indian perspective. na, 1981, p.69.

Silverman, David J. Native American Religions. Oxford University Press, 2013.

Sheldrake, Philip, The Spiritual City: Theology, Spirituality, and the Urban, First Edition. John Wiley & Sons, Ltd., 2014, p. 117-118.

Sluyter, Ander (2003). "Neo-Environmental Determinism, Intellectual Damage Control, and Nature/Society Science". Antipode-Blackwell.

Smith, Huston. A seat at the table: Huston Smith in conversation with Native Americans on religious freedom. Univ of California Press, 2007, p. 39-57.

Standing Bear, Luther, 2006, Land of the Spotted Eagle. Lincoln: Univ. Nebraska Press.

Stausberg, M., Religion and tourism: Crossroads, destinations and encounters. Routledge, 2012.

Sturtevant, William C. Handbook of North American Indians. Eds. Wilcomb E. Washburn, et al. Vol. 8. Washington, DC: Smithsonian Institution, 1978.

Timothy, Dallen J., and Paul J. Conover. "10 Nature religion, self-spirituality and New Age tourism." Tourism, religion and spiritual journeys (2006): 139-149.

Waldman, Carl, and Braun, Molly. Atlas of the North American Indian. Infobase, 2009.

Walker, James R. Lakota belief and ritual. U of Nebraska Press, 1980.

Walker, J., (1982) Lakota Society, ed. Raymond J. DeMallie. Lincoln: Univ. of Nebraska Press.

Walker, James R. Lakota myth. U of Nebraska Press, 2006, p. 10-12.

Warner, Elizabeth Ann Kronk. "Looking to the third sovereign: Tribal environmental ethics as an alternative paradigm." Pace Envtl. L. Rev. 33 (2015): 397.

Welch, Christina. "Appropriating the didjeridu and the sweat lodge: new age baddies and Indigenous victims?." Journal of Contemporary Religion 17.1 (2002): 21-38.

White Lance, Frank, Why the Black Hills are Sacred: A Unified Theory of the Lakota Sundance, Ancestors Inc, Rapid City, SD, 2004

Whyte, Kyle. "The Dakota access pipeline, environmental injustice, and US colonialism." Red Ink: An International Journal of Indigenous Literature, Arts, & Humanities 19.1 (2017).

Wishart, David, J., Editor, Encyclopedia of the Great Plains Indians, Univ. of Nebraska, 2004.

Wuthnow, Robert. "Creative spirituality: the way of the artist." Nova Religio (2005): 123-125.

Zimmerman, Larry J., and Brian Molyneaux. Native North America. Norman, OK: University of Oklahoma Press, 1996.

B. Internet Website Sources and Personal Communications

1. Akta Lakota Museum and Cultural Center, "Sioux" Name and Dialects, at: http://aktalakota.stjo.org/site/PageServer?pagename=alm_culture_or igins, accessed August 30, 2019.

2. Badlands National Park, Geologic Formations: https://www.nps.gov/badl/learn/nature/geologicformations.htm, accessed September 21, 2019.

3. Bidwell, Laural, A., Geography and Geology of the Black Hills: https://www.moon.com/travel/trip-ideas/geography-geology-black-hills/, accessed September 6, 2019.

4. Dr. Joan Borysenko's website: https://www.joanborysenko.com/spirituality/what-is-interspirituality/. Accessed January 6, 2020.

5. How Did the Badlands National Park Get Its Name?: https://www.blackhillsbadlands.com/blog/2014-07-10/how-did-badlands-national-park-get-its-name, accessed September 21, 2019.

6. My Tribal Area, U.S. Census Bureau at: https://www.census.gov/tribal/?aianihh=2810, accessed September 8, 2019.

7. Pine Ridge Indian Reservation, Re-Member website: https://www.re-member.org/pine-ridge-reservation.aspx, accessed September 8, 2019

8. The Medicine Wheel, St. Joseph's Indian School website: https://www.stjo.org/native-american-culture/native-american-beliefs/medicine-wheel/, accessed September 28, 2019.

9. Akta Lakota Museum and Cultural Center, The Four Directions, at: http://aktalakota.stjo.org/site/News2?page=NewsArticle&id=8593, accessed September 4, 2019

10. Bamfroth, Douglas, B., Intertribal Warfare, Encyclopedia of the Great Plains, University of Nebraska Lincoln, at: http://plainshumanities.unl.edu/encyclopedia/doc/egp.war.023, accessed August 31, 2019.

11. Black Hills and Badlands Geology: https://blackhillsvisitor.com/learn/black-hills-and-badlands-geology/2/, accessed September 6, 2019.

12. Black Hills in South Dakota and Wyoming Map, Raremaps.com at: https://www.raremaps.com/gallery/detail/38609/map-of-the-black-hills-of-south-dakota-and-wyoming-with-pam-scottaccessed September 6, 2019

13. Clark, Linda, Darus, Sioux Treaty of 1868, National Archives, at: https://www.archives.gov/education/lessons/sioux-treaty, accessed August 31, 2019.

14. Clinton, William, J., President, Indian Sacred Sites Executive Order, May 24, 1996 at: https://www.nps.gov/history/local-law/eo13007.htm, accessed September 2, 2019.

15. Conceptually (January 20, 2019). "Determinism - Explanation and examples". conceptually.org, accessed September 27, 2019.

16. Giese, Paula Lakota Star Knowledge, 1995 at: http://www.kstrom.net/isk/stars/startabs.html, accessed September 29, 2019.

17. Oceti Sakowin, Seven Council Fires, Akta Lakota Museum and Cultural Center website: http://aktalakota.stjo.org/site/News2?page=NewsArticle&id=8309, accessed August 30, 2019

18. Personal communication on September 12, 2019 with Jace DeCory, an enrolled member of the Cheyenne River Lakota Nation, and Professor Emeritus, Black Hills State University.

19. Personal communication with David Posthumus on September 4, 2019 and personal communication with Jace DeCory on September 12, 2019.

20. Personal communication with Jace DeCory in Spearfish, SD on September 12, 2019.

21. Personal communication with Professor Jace DeCory on September 12, 2019 in Spearfish, SD.

22. Personal communication with Roger Broer, a prominent artist, Lakota elder and enrolled member of the Oglala Lakota Nation on September 3, 2019.

23. Personal communications with Jace DeCory (Cheyenne River Lakota) at Black Hills State University on September 12, 2019 and Roger Broer (Oglala Lakota) on September 3, 2019.

24. Pew Research Center, Religious Landscape Study at: https://www.pewforum.org/religious-landscape-study/#religions, accessed September 25, 2019.

25. Pine Ridge Indian Reservation, Re-Member website: https://www.re-member.org/pine-ridge-reservation.aspx, accessed September 8, 2019.

26. Sacred Land Film Project, what is a Sacred Site? Website: https://sacredland.org/tools-for-action/. Accessed September 4, 2019.

27. Simmins, Geoffrey, Sacred Spaces and Sacred Places, Book draft accessed at: https://dspace.ucalgary.ca/bitstream/handle/1880/46834/Sacred%20Spaces.pdf;jsessionid=B2B030886651BE31E5E5349214B504F6?sequence=1 on September 4, 2019

28. Sioux Nation Treaty Council, Great Sioux Reservation at: https://www.siouxnationtreatycouncil.org/index.php/maps/26-great-sioux-reservation, accessed August 30, 2019.

29. Smithsonian Institutions, Museum of the American Indian, Native Knowledge 360 ° Framework for Essential Understandings about American Indians at: https://americanindian.si.edu/nk360/pdf/NMAI-Essential-Understandings.pdf, accessed September 2, 2019.

30. Treaty of Fort Laramie, 1851, at: https://www.ourdocuments.gov/doc.php?flash=false&doc=42, accessed on August 30, 2019

In Sacred Relationship

APPENDIX 1:
Pine Ridge Reservation, South Dakota/Nebraska
Demographic, Employment, and Educational Profile
Source: 2014-2018 American Community Survey 5-Year Estimates

Subject	Title	Pine Ridge Reservation SD--NE
Sex and Age	Total population	19,895
Sex and Age	Male	9,699
Sex and Age	Female	10,196
Sex and Age	Under 5 years	2,045
Sex and Age	5 to 9 years	1,937
Sex and Age	10 to 14 years	2,259
Sex and Age	15 to 19 years	1,768
Sex and Age	20 to 24 years	1,645
Sex and Age	25 to 34 years	2,794
Sex and Age	35 to 44 years	2,049
Sex and Age	45 to 54 years	2,110
Sex and Age	55 to 59 years	944
Sex and Age	60 to 64 years	778
Sex and Age	65 to 74 years	913
Sex and Age	75 to 84 years	551
Sex and Age	85 years and over	102
Sex and Age	Median age (years)	26
Sex and Age	18 years and over	12,561
Sex and Age	65 years and over	1,566
Race	Total population	19,895
Race	One race	19,145
Race	White	2,351
Race	Black or African American	17
Race	American Indian and Alaska Native	16,714
Race	Asian	26
Race	Native Hawaiian and Other Pacific Islander	-

In Sacred Relationship

Subject	Title	Pine Ridge Reservation SD--NE
Race	Some other race	37
Race	Two or more races	750
Hispanic or Latino and Race	Total population	19,895
Hispanic or Latino and Race	Hispanic or Latino (of any race)	867
Hispanic or Latino and Race	Mexican	788
Hispanic or Latino and Race	Puerto Rican	21
Hispanic or Latino and Race	Cuban	-
Hispanic or Latino and Race	Other Hispanic or Latino	58
Hispanic or Latino and Race	Not Hispanic or Latino	19,028
Place of Birth	Total population	19,895
Place of Birth	Native	19,814
Place of Birth	Born in United States	19,771
Place of Birth	State of residence	16,482
Place of Birth	Different state	3,289
Place of Birth	Born in Puerto Rico, U.S. Island areas, or born abroad to American parent(s)	43
Place of Birth	Foreign born	81
Ancestry	Total population	19,895
Ancestry	American	153
Ancestry	Arab	-
Ancestry	Czech	14
Ancestry	Danish	38
Ancestry	Dutch	49
Ancestry	English	175
Ancestry	French (except Basque)	680
Ancestry	French Canadian	23
Ancestry	German	1,272

In Sacred Relationship

Subject	Title	Pine Ridge Reservation SD--NE
Ancestry	Greek	-
Ancestry	Hungarian	-
Ancestry	Irish	410
Ancestry	Italian	19
Ancestry	Lithuanian	-
Ancestry	Norwegian	222
Ancestry	Polish	57
Ancestry	Portuguese	-
Ancestry	Russian	61
Ancestry	Scotch-Irish	32
Ancestry	Scottish	49
Ancestry	Slovak	-
Ancestry	Sub-Saharan African	3
Ancestry	Swedish	122
Ancestry	Swiss	6
Ancestry	Ukrainian	-
Ancestry	Welsh	12
Ancestry	West Indian (excluding Hispanic origin groups)	5
Veteran Status	Civilian population 18 years and over	12,561
Veteran Status	Civilian veterans	840
Disability Status of the Civilian Noninstitutionalized Population	Total civilian noninstitutionalized population	19,836
Disability Status of the Civilian Noninstitutionalized Population	With a disability	2,759
Disability Status of the Civilian Noninstitutionalized Population	Under 18 years	7,334
Disability Status of the Civilian	With a disability	339

In Sacred Relationship

Subject	Title	Pine Ridge Reservation SD--NE
Noninstitutionalized Population		
Disability Status of the Civilian Noninstitutionalized Population	18 to 64 years	10,981
Disability Status of the Civilian Noninstitutionalized Population	With a disability	1,677
Disability Status of the Civilian Noninstitutionalized Population	65 years and over	1,521
Disability Status of the Civilian Noninstitutionalized Population	With a disability	743
Residence 1 Year Ago	Population 1 year and over	19,526
Residence 1 Year Ago	Same house	18,503
Residence 1 Year Ago	Different house in the U.S.	1,016
Residence 1 Year Ago	Same county	459
Residence 1 Year Ago	Different county	557
Residence 1 Year Ago	Same state	345
Residence 1 Year Ago	Different state	212
Residence 1 Year Ago	Abroad	7
Employment Status	Population 16 years and over	13,395
Employment Status	In labor force	6,263
Employment Status	Civilian labor force	6,263
Employment Status	Employed	5,179
Employment Status	Unemployed	1,084
Employment Status	Armed Forces	-
Employment Status	Not in labor force	7,132

In Sacred Relationship

Subject	Title	Pine Ridge Reservation SD--NE
Employment Status	Civilian labor force	6,263
Employment Status	Unemployment Rate	17
Commuting to Work	Workers 16 years and over	4,957
Commuting to Work	Car, truck, or van -- drove alone	3,214
Commuting to Work	Car, truck, or van -- carpooled	498
Commuting to Work	Public transportation (excluding taxicab)	113
Commuting to Work	Walked	567
Commuting to Work	Other means	233
Commuting to Work	Worked at home	332
Commuting to Work	Mean travel time to work (minutes)	18
Occupation	Civilian employed population 16 years and over	5,179
Occupation	Management, business, science, and arts occupations	1,892
Occupation	Service occupations	1,393
Occupation	Sales and office occupations	834
Occupation	Natural resources, construction, and maintenance occupations	524
Occupation	Production, transportation, and material moving occupations	536
Industry	Civilian employed population 16 years and over	5,179
Industry	Agriculture, forestry, fishing and hunting, and mining	451
Industry	Construction	181
Industry	Manufacturing	100

In Sacred Relationship

Subject	Title	Pine Ridge Reservation SD--NE
Industry	Wholesale trade	67
Industry	Retail trade	445
Industry	Transportation and warehousing, and utilities	106
Industry	Information	47
Industry	Finance and insurance, and real estate and rental and leasing	144
Industry	Professional, scientific, and management, and administrative and waste management services	118
Industry	Educational services, and health care and social assistance	1,995
Industry	Arts, entertainment, and recreation, and accommodation and food services	583
Industry	Other services, except public administration	138
Industry	Public administration	804
Class of Worker	Civilian employed population 16 years and over	5,179
Class of Worker	Private wage and salary workers	1,856
Class of Worker	Government workers	2,837
Class of Worker	Self-employed in own not incorporated business workers	459
Class of Worker	Unpaid family workers	27
Housing Occupancy	Total housing units	5,413
Housing Occupancy	Occupied housing units	4,246
Housing Occupancy	Vacant housing units	1,167
Housing Occupancy	Homeowner vacancy rate	0
Housing Occupancy	Rental vacancy rate	4

In Sacred Relationship

Subject	Title	Pine Ridge Reservation SD--NE
Housing Tenure	Occupied housing units	4,246
Housing Tenure	Owner-occupied	2,313
Housing Tenure	Renter-occupied	1,933
Housing Tenure	Average household size of owner-occupied unit	4
Housing Tenure	Average household size of renter-occupied unit	5
Year Householder Moved into Unit	Occupied housing units	4,246
Year Householder Moved into Unit	Moved in 2017 or later	77
Year Householder Moved into Unit	Moved in 2015 to 2016	252
Year Householder Moved into Unit	Moved in 2010 to 2014	727
Year Householder Moved into Unit	Moved in 2000 to 2009	1,191
Year Householder Moved into Unit	Moved in 1990 to 1999	1,099
Year Householder Moved into Unit	Moved in 1989 and earlier	900
Value	Owner-occupied units	2,313
Value	Less than $50,000	1,189
Value	$50,000 to $99,999	630
Value	$100,000 to $149,999	226
Value	$150,000 to $199,999	77
Value	$200,000 to $299,999	112
Value	$300,000 to $499,999	29
Value	$500,000 to $999,999	29
Value	$1,000,000 or more	21
Value	Median (dollars)	47,000
Mortgage Status	Owner-occupied units	2,313
Mortgage Status	Housing units with a mortgage	445
Mortgage Status	Housing units without a mortgage	1,868

In Sacred Relationship

Subject	Title	Pine Ridge Reservation SD--NE
Selected Monthly Owner Costs (SMOC)	Housing units with a mortgage	445
Selected Monthly Owner Costs (SMOC)	Less than $500	36
Selected Monthly Owner Costs (SMOC)	$500 to $999	207
Selected Monthly Owner Costs (SMOC)	$1,000 to $1,499	158
Selected Monthly Owner Costs (SMOC)	$1,500 to $1,999	34
Selected Monthly Owner Costs (SMOC)	$2,000 to $2,499	-
Selected Monthly Owner Costs (SMOC)	$2,500 to $2,999	10
Selected Monthly Owner Costs (SMOC)	$3,000 or more	-
Selected Monthly Owner Costs (SMOC)	Median (dollars)	921
Selected Monthly Owner Costs (SMOC)	Housing units without a mortgage	1,868
Selected Monthly Owner Costs (SMOC)	Less than $250	408
Selected Monthly Owner Costs (SMOC)	$250 to $399	634
Selected Monthly Owner Costs (SMOC)	$400 to $599	560
Selected Monthly Owner Costs (SMOC)	$600 to $799	194

In Sacred Relationship

Subject	Title	Pine Ridge Reservation SD--NE
Selected Monthly Owner Costs (SMOC)	$800 to $999	36
Selected Monthly Owner Costs (SMOC)	$1,000 or more	36
Selected Monthly Owner Costs (SMOC)	Median (dollars)	378
Gross Rent	Occupied units paying rent	1,719
Gross Rent	Less than $500	760
Gross Rent	$500 to $999	893
Gross Rent	$1,000 to $1,499	58
Gross Rent	$1,500 to $1,999	8
Gross Rent	$2,000 to $2,499	-
Gross Rent	$2,500 to $2,999	-
Gross Rent	$3,000 or more	-
Gross Rent	Median (dollars)	529
Gross Rent	No rent paid	214
Computer and Internet Use	Total Households	4,246
Computer and Internet Use	Percent of households with a computer	63
Computer and Internet Use	Percent of households with a broadband Internet subscription	53
Income and Benefits (In 2018 inflation-adjusted dollars)	Total households	4,246
Income and Benefits (In 2018 inflation-adjusted dollars)	Less than $10,000	779
Income and Benefits (In 2018 inflation-adjusted dollars)	$10,000 to $14,999	388
Income and Benefits (In 2018 inflation-adjusted dollars)	$15,000 to $24,999	474

In Sacred Relationship

Subject	Title	Pine Ridge Reservation SD--NE
Income and Benefits (In 2018 inflation-adjusted dollars)	$25,000 to $34,999	514
Income and Benefits (In 2018 inflation-adjusted dollars)	$35,000 to $49,999	603
Income and Benefits (In 2018 inflation-adjusted dollars)	$50,000 to $74,999	691
Income and Benefits (In 2018 inflation-adjusted dollars)	$75,000 to $99,999	378
Income and Benefits (In 2018 inflation-adjusted dollars)	$100,000 to $149,999	276
Income and Benefits (In 2018 inflation-adjusted dollars)	$150,000 to $199,000	53
Income and Benefits (In 2018 inflation-adjusted dollars)	$200,000 or more	90
Income and Benefits (In 2018 inflation-adjusted dollars)	Median household income (dollars)	34,380
Income and Benefits (In 2018 inflation-adjusted dollars)	Mean household income (dollars)	45,982
Health Insurance Coverage	Civilian noninstitutionalized population	19,836
Health Insurance Coverage	With health insurance coverage	13,425
Health Insurance Coverage	With private health insurance	3,937
Health Insurance Coverage	With public coverage	10,265
Health Insurance Coverage	No health insurance coverage	6,411
Health Insurance Coverage	Civilian noninstitutionalized population under 19 years	7,625
Health Insurance Coverage	No health insurance coverage	1,227

In Sacred Relationship

Subject	Title	Pine Ridge Reservation SD--NE
Percentage of Families and People Whose Income in the Past 12 Months is Below the Poverty Level	All families	40
Percentage of Families and People Whose Income in the Past 12 Months is Below the Poverty Level	With related children of the householder under 18 years	49
Percentage of Families and People Whose Income in the Past 12 Months is Below the Poverty Level	With related children of the householder under 5 years only	48
Percentage of Families and People Whose Income in the Past 12 Months is Below the Poverty Level	Married couple families	20
Percentage of Families and People Whose Income in the Past 12 Months is Below the Poverty Level	With related children of the householder under 18 years	26
Percentage of Families and People Whose Income in the Past 12 Months is Below the Poverty Level	With related children of the householder under 5 years only	22
Percentage of Families and People Whose Income in the Past 12 Months is Below the Poverty Level	Families with female householder, no husband present	55
Percentage of Families and People Whose Income in	With related children of the householder under 18 years	64

In Sacred Relationship

Subject	Title	Pine Ridge Reservation SD--NE
the Past 12 Months is Below the Poverty Level		
Percentage of Families and People Whose Income in the Past 12 Months is Below the Poverty Level	With related children of the householder under 5 years only	49
Percentage of Families and People Whose Income in the Past 12 Months is Below the Poverty Level	All people	48
Percentage of Families and People Whose Income in the Past 12 Months is Below the Poverty Level	Under 18 years	58
Percentage of Families and People Whose Income in the Past 12 Months is Below the Poverty Level	Related children of the householder under 18 years	58
Percentage of Families and People Whose Income in the Past 12 Months is Below the Poverty Level	Related children of the householder under 5 years	60
Percentage of Families and People Whose Income in the Past 12 Months is Below the Poverty Level	Related children of the householder 5 to 17 years	56
Percentage of Families and People Whose Income in the Past 12 Months is Below the Poverty Level	18 years and over	43

In Sacred Relationship

Subject	Title	Pine Ridge Reservation SD--NE
Percentage of Families and People Whose Income in the Past 12 Months is Below the Poverty Level	65 years and over	30
Percentage of Families and People Whose Income in the Past 12 Months is Below the Poverty Level	People in families	47
Percentage of Families and People Whose Income in the Past 12 Months is Below the Poverty Level	Unrelated individuals 15 years and over	57
School Enrollment	Population 3 years and over enrolled in school	7,204
School Enrollment	Nursery school, preschool	607
School Enrollment	Kindergarten	444
School Enrollment	Elementary school (grades 1-8)	3,361
School Enrollment	High school (grades 9-12)	1,532
School Enrollment	College or graduate school	1,260
Educational Attainment	Population 25 years and over	10,241
Educational Attainment	Less than 9th grade	480
Educational Attainment	9th to 12th grade, no diploma	1,845
Educational Attainment	High school graduate (includes equivalency)	2,797
Educational Attainment	Some college, no degree	2,865
Educational Attainment	Associate degree	878
Educational Attainment	Bachelor's degree	1,062

In Sacred Relationship

Subject	Title	Pine Ridge Reservation SD--NE
Educational Attainment	Graduate or professional degree	314
Educational Attainment	Percent high school graduate or higher	77
Educational Attainment	Percent bachelor's degree or higher	13

In Sacred Relationship

APPENDIX 2:
SOUTH DAKOTA TRIBAL DEMOGRAPHIC COMPARISON
Source: U.S. Census Tribal Area Data

Variable	Pine Ridge	Cheyenne River	Flandreau	Lake Traverse	Lower Brule	Rosebud	Standing Rock	Yankton	Crow Creek
Population by Age in 2017									
Total Population	19,779	8,527	442	10,967	1,594	11,354	8,616	6,676	2,151
Preschool (0 to 4)	2,080	885	36	981	194	1,402	908	675	243
School Age (5 to 17)	5,188	2,107	103	2,240	388	3,102	2,223	1,479	603
College Age (18 to 24)	2,341	884	32	879	212	1,242	843	648	249
Young Adult (25 to 44)	4,824	1,996	87	2,225	420	2,730	2,006	1,327	487
Adult (45 to 64)	3,872	1,863	154	2,791	286	2,105	1,874	1,460	423
Older Adult (65 plus)	1,474	792	30	1,851	94	773	762	1,087	146
Population by Race in 2017									
Total Population	19,779	8,527	442	10,967	1,594	11,354	8,616	6,676	2,151
American Ind. or Alaskan Native Alone	16,501	6,554	328	4,363	1,477	9,072	6,538	2,935	1,943
Asian Alone	25	33	0	4	3	131	11	21	0
Black Alone	29	8	4	48	2	32	21	4	8
Native Hawaiian and Other Pac. Isl. Alone	0	0	0	0	3	0	2	10	40
White Alone	2,224	1,815	31	5,972	95	943	1,807	3,365	112
Two or More Race Groups	964	113	75	376	14	1,160	204	319	45
Total Hispanic or Latino	805	86	30	637	31	505	204	286	71
Mexican	773	47	16	469	16	426	154	195	71
Cuban	0	0	0	1	0	17	23	33	0
Puerto Rican	3	4	0	25	0	22	0	24	0
Other	29	35	14	142	15	40	27	34	0
Educational Attainment in 2017									
Population 25 and Older	10,170	4,651	271	6,867	800	5,608	4,642	3,874	1,056
Less Than 9th Grade	548	176	13	335	44	287	157	177	65

In Sacred Relationship

Variable	Pine Ridge	Cheyenne River	Flandreau	Lake Traverse	Lower Brule	Rosebud	Standing Rock	Yankton	Crow Creek
9th to 12th, No Diploma	1,808	560	22	522	116	1,027	584	427	154
High School Graduate (incl. equiv.)	2,748	1,737	78	2,435	258	1,645	1,531	1,386	434
Some College, No Degree	2,888	964	79	1,415	228	1,511	997	751	250
Associate Degree	790	531	32	959	91	301	559	463	70
Bachelor's Degree	1,086	519	35	793	53	553	642	469	63
Graduate or Professional Degree	302	164	12	408	10	284	172	201	20
Households in 2017									
Total Households	4,350	2,403	190	3,960	402	3,148	2,362	2,102	535
Family Households	3,474	1,721	122	2,730	335	2,293	1,780	1,361	433
Married with Children	545	384	27	570	45	370	318	291	91
Married without Children	789	584	40	1,162	83	503	544	595	90
Single Parents	984	355	31	573	130	834	449	255	110
Other	1,156	398	24	425	77	586	469	220	142
Non-family Households	876	682	68	1,230	67	855	582	741	102
Living Alone	779	582	56	1,101	55	732	488	633	95
Housing Units in 2017									
Total Housing Units	5,374	3,010	206	5,641	453	3,610	2,896	2,578	647
Owner Occupied	2,341	1,365	79	2,497	160	1,317	1,160	1,354	208
Renter Occupied	2,009	1,038	111	1,463	242	1,831	1,202	748	327
Vacant for Seasonal or Recreational Use	1,024	607	16	1,681	51	462	534	476	112
1-Unit (Attached or Detached)	2,966	1,614	133	3,372	335	2,116	1,876	1,729	421
2 - 9 Units	138	171	35	200	39	294	128	133	9
10 - 19 Units	64	34	9	90	0	97	36	62	21
20 or more Units	0	59	9	66	3	79	0	38	0
Built prior to 1940	323	283	4	1,391	17	160	330	693	33
Commuting to Work in 2017									

In Sacred Relationship

Variable	Pine Ridge	Cheyenne River	Flandreau	Lake Traverse	Lower Brule	Rosebud	Standing Rock	Yankton	Crow Creek
Workers 16 years and over	4,943	2,903	169	4,825	505	3,074	2,606	2,430	601
Car, truck, or van -- drove alone	2,969	1,999	131	3,381	355	2,075	1,687	1,803	429
Car, truck, or van -- carpooled	560	157	6	610	69	344	262	192	68
Public transportation (including taxicab)	93	8	0	37	8	29	14	10	3
Walked	522	95	22	249	22	171	142	117	51
Other means	213	47	7	58	0	102	27	65	9
Worked at home	391	532	0	330	19	297	429	197	23
Resident Occupations in 2017									
Employed civilian pop. 16 years and over	6,192	3,943	186	5,220	600	3,640	3,429	2,622	732
Management, professional, and related	1,761	1,331	47	1,929	205	1,130	1,089	934	175
Service	1,325	485	61	911	81	730	548	560	182
Sales and office	890	539	32	755	114	637	500	487	136
Farming, fishing, and forestry	108	123	1	230	8	151	96	26	15
Construction, extraction, and maintenance	359	259	18	342	61	213	168	154	45
Production, transportation, and material moving	500	166	10	658	36	213	205	269	48

INDEX

Acamo Pueblo, 47
All Lives Matter, 121
all my relations, 41, 44, 85
Allegheny Health Network, 152
American Indian Country, 27, 51
Ancestral Spiritual Ties, 73
Animism, 71, 72, 154
Anthropology, 1, 8, 14, 15, 18, 152
As Above, So Below, 74
Auric field, 97
Authenticity, 117
Badlands, iii, 12, 16, 17, 28, 43, 48, 52, 53, 59, 60, 61, 62, 63, 71, 72, 127, 152, 159, 160
Badlands Geological History, 60
Bear Butte, 58, 67
Belmont County, Ohio, 14, 15
Bighorn Sheep, 62, 63
Bison, 28, 34, 41, 58, 59, 67, 73, 156
Black Elk, 10, 27, 42, 43, 52, 152
Black Hills, 12
Black Hills Geological History, 53
Black Hills National Park, 52
C. G. Glacken, 70
Cancer, iii, 92, 94, 130, 131, 132, 133, 135, 136, 137, 138, 140, 141, 142, 151, 154
Cancer Care, iii, 130, 140
Carl Sagan, 9
Chakras, 99, 101, 102, 104, 105
Charismatic Movement, 35
Charles Foster, 46
Cherokee Nation of Oklahoma, 16, 152
Chief Seattle, 2, 5
Christianity, 12, 14, 34, 35, 40, 95
Cleveland Clinic, 1, 94, 96, 134, 151
Cleveland State University, 15, 16, 151
complementary medicine, 94, 131, 151

Core Sacred Relationship Types, 115
Coronavirus Pandemic, 6, 126, 129, 143, 144, 145, 146, 147
cosmology, 57, 58
COVID-19, 143, 144, 147
Crazy Horse, 27, 28, 43, 55, 67
Dakota pipeline, 49
David Posthumus, 1, 36, 42, 72, 73, 161
David Wishart, 46
Dawes Act of 1887, 56
Deepak Chopra, 99
Depth psychology, 79
Devils Tower, 12
Devils Tower Geological History, 64
Divinity, 9, 16, 152
Don Iannone, 3, 54, 56, 59, 61, 63, 65, 66, 151
Douglas McKenzie, 16
Earth Spirituality, 37, 40, 155
Ed McGaa, 40
energy medicine, 92, 95, 96, 97, 98, 99, 100, 101, 105
Energy medicine, 95, 96
Eva Marie Garroutte, 35
Fools Crow, 42
Four Types of Core Sacred Relationships, 114
Gallup, 5
Geoffrey Simmins, 47
Ghost Dance, 55, 56, 61, 62, 153
Great Mystery, 41, 67, 69, 72
Great Spirit, 36, 38, 41, 43, 60, 62, 75, 76, 79
Grounded spirituality, 10, 80
grounded spirituality and sacred relationship model, 81, 124
Higher Spiritual Good, 145, 147, 148, 149
Human Energy Body, 95
In Sacred Relationship, 2, iii, 1, 4, 5, 8, 9, 14, 17, 37, 126
Integrated health and medicine, 93
Interfaith spirituality, 78
Interspirituality, 35, 78
Itokaga, 40
Jace DeCory, 1, 31, 32, 72, 73, 161
James Hillman, 79
Jeffrey Ostler, 66
Jesus Christ, 12
Jill Oakes, 45

Joan Borysenko, 159

Joel Brady, 50

John Blank, 16

Lakota, i, 2, iii, 1, 3, 7, 8, 9, 10, 11, 14, 16, 17, 18, 19, 20, 21, 27, 28, 29, 30, 31, 33, 36, 37, 38, 39, 40, 41, 42, 43, 44, 45, 46, 48, 49, 51, 52, 54, 55, 56, 57, 58, 59, 60, 61, 62, 64, 66, 67, 68, 69, 70, 71, 72, 73, 74, 75, 76, 78, 79, 85, 127, 152, 153, 154, 155, 156, 158, 159, 160, 161

Larry Zimmerman, 1

Lisa A. McLoughlin, 44

Luther Standing Bear, 21, 72, 78, 129

Maka Wita, 57

Martins Ferry, 152

Mato Tipila, 64

Medicine Wheel, 39, 159

Meridians, 99

Mikael Kurkiala, 28

Mind-Body Medicine, 152

Moral Stewardship, 73

Morphological field, 97

Mother Earth, 37, 40, 45, 71, 155

Mysticism, 36

Native Americans, 22, 23, 24, 25, 28, 31, 34, 37, 38, 39, 47, 49, 50, 55, 57, 60, 64, 74, 157

Nazarene Church, 12, 14

Negative Emotions, 117

Oceti Sakowin, 30, 33, 41, 54, 160

Oglala Lakota, 1, 11, 41, 42, 55, 57, 62, 72, 161

Ohio Historical Society, 15

Oneness, 73, 91, 116

Pahá Sápa, 58

Pathways to Sacred Relationships, 119, 120

Personal branding, 119

Peter Nabokov, 44

Philip Sheldrake, 70

Physical (Material) Body, 90, 103

Pine Ridge, 11, 25, 26, 28, 33, 34, 55, 56, 57, 61, 62, 159, 161, 162, 176

Planet Earth, 123, 124, 125, 126

Population, 25, 124, 165, 166, 174, 175, 176, 177

Posthumus, David, 184

President Trump, 24

Raul M. Grijalva, 23

Reiki, 96, 104, 152

Religion and Spirituality, 133
Roger Broar, 1
Rudolph Otto, 46
Sacralization, 45, 75, 127
Sacred Land, iii, 44, 48, 60, 69, 127, 161
Sacred Meaning, 45, 74
Sacred Relationship Toxicity, 117
Sacred Relationships with Others, 113
Sacred Relationships with the Earth, 124
Sacred Relationships with Your Body, 89
Seven Council Fires, 11, 33, 41, 54, 160
Soul Proximity, 120
South Dakota, iii, 1, 11, 12, 16, 17, 21, 22, 25, 26, 27, 28, 29, 33, 37, 42, 43, 48, 53, 54, 55, 57, 58, 59, 61, 63, 67, 68, 69, 157, 160, 162
spiritual care, 94, 134
Spiritual Compass, i, 81, 147, 150
St. Clairsville, 152
Sun dance, 47
Ten Commandments, 80, 81
Tetonwan, 11, 33
Tianjun Liu, 96
Traditional Chinese Medicine, 95, 96, 99
Trump Administration, 24
Vine Deloria, 20, 44, 48
Wakan Tanka, 38, 41, 43, 69, 72
Wayne Teasdale, 35
Waziyata, 40
While Buffalo Calf Woman, 67
White Buffalo Calf Pipe, 67
Wholeness, 121
William K. Powers, 33, 41, 58
Wind Cave, 58, 68
Winona LaDuke, 71
Wioheumpata, 39
Wiyokpiyata, 40
Worldview, 10, 11, 12, 14, 31, 39, 40, 42, 44, 58, 69, 70, 72, 73, 74, 127
Wounded Knee, 10, 28, 55, 56, 62, 154
Zimmerman and Molyneaux, 42, 74

CPSIA information can be obtained
at www.ICGtesting.com
Printed in the USA
FSHW021529050420
68835FS